Advances in Computer Vision and Pattern Recognition



For other titles published in this series, go to
www.springer.com/series/4205

Cem Ünsalan • Kim L. Boyer

# Multispectral Satellite Image Understanding

## From Land Classification to Building and Road Detection

Springer

Assoc. Prof. Cem Ünsalan
Electrical and Electronics Engineering
Yeditepe University
26 Ağustos Yerleşimi
34755 Kayisdagi, Istanbul
Turkey
unsalan@yeditepe.edu.tr

Prof. Kim L. Boyer
Dept. Electrical, Comp. & Systems Eng.
Rensselaer Polytechnic Institute
8th Street 110
12180 Troy, NY
USA
kim@ecse.rpi.edu

*Series Editors*
Professor Sameer Singh, PhD
Research School of Informatics
Loughborough University
Loughborough
UK

Dr. Sing Bing Kang
Microsoft Research
Microsoft Corporation
One Microsoft Way
Redmond, WA 98052
USA

ISSN 2191-6586
e-ISSN 2191-6594
ISBN 978-1-4471-2656-0
ISBN 978-0-85729-667-2 (eBook)
DOI 10.1007/978-0-85729-667-2
Springer London Dordrecht Heidelberg New York

British Library Cataloguing in Publication Data
A catalogue record for this book is available from the British Library



*Cover design*: VTeX UAB, Lithuania

Printed on acid-free paper

Springer is part of Springer Science+Business Media (www.springer.com)

*To our families.*

# Foreword

In the first decade of the twenty first century, remote sensing has undergone a rapid development, boosting many new or improved application possibilities. This is due to higher spatial resolution of satellite image data as well as better data availability regarding quality, frequency and coverage. On the other hand, the scientific development of image and signal processing has lead to more powerful and reliable methods, which in turn result in better and faster evaluation of the huge amounts of data sets using fully automatic procedures. The book at hand contributes to this development by combining methods from image processing and electrical engineering stimulated by computer science and computer vision technologies.

There have been many publications in journals and also books on the mentioned topics, but in most cases they show certain specializations either on theory or on applications. The special value of this book is that it presents a complete chain of image processing methods to derive reliable information for land use, especially in residential areas. The authors know very well how to combine a theoretical framework like graph formalism with very practical applications. They use well known methods (e.g., NDVI) together with new techniques from computer vision to arrive at a system which allows detecting single objects like houses and streets in very high resolution optical images (e.g., IKONOS) effectively. The presented system can be applied for change detection as well as other quantitative analysis of urban development.

Due to the fast growth of the remote sensing market, automatic image processing methods exhibit an increasing potential for more and more applications. Through tailoring the described methods for fitting his task, the reader will be able to set up his own system to extract the desired information or develop new methods based on the given techniques. Therefore, I hope the book will be a further milestone from scientific remote sensing to practical applications.

Wessling, Germany Prof. Dr. Peter Reinartz

# Preface

As the resolution of satellite images increased, more detailed analysis on them became possible. On the other hand, the time required to manually analyze them became prohibitive. Hence, the need for automated systems for such analysis tasks emerged. This book is about such an end-to-end image analysis system to understand land development from satellite images. Our focus is on residential regions. The main building blocks of the proposed system are as follows.

We benefit from vegetation and shadow–water indices in summarizing the multispectral information in the proposed system. Vegetation indices have been used extensively to estimate the vegetation density from satellite and airborne images for many years. We focus on the normalized difference vegetation index (*NDVI*) and introduce a statistical framework to analyze and extend it. Using the established statistical framework, we introduce new a group of shadow–water indices. We use these as the source of multispectral information in land use classification and house and street network detection in residential regions.

Next, we introduce a set of measures based on straight lines to assess land development levels in high resolution satellite images. Urban areas exhibit a preponderance of straight line features. Rural areas produce line structures in more random spatial arrangements. We use this observation to perform an initial triage on the image to restrict the attention of subsequent, more computationally intensive analyses. We then extend our straight line based measures by developing a synergistic approach that combines structural and multispectral information. In particular, the structural features serve as cue regions for multispectral features.

After the initial classification of regions, we introduce computationally more expensive but more precise graph-theoretical measures over panchromatic images to detect residential regions. The graphs are constructed using straight lines as vertices, while graph edges encode their spatial relationships. We introduce a set of measures based on various properties of the graph. These measures are monotonic with increasing structure (organization) in the image. We present a theoretical basis for the measures. In a similar manner, we developed a novel method using feature based grouping to detect residential regions.

Having detected the residential regions, we introduce a novel subsystem to detect houses and street networks in these. This system is composed of four main blocks:

detecting possible house and street pixels by the help of multispectral information; grouping these candidate pixels using a variant of k-means clustering; decomposing the clustering results by a novel balloon algorithm; and finally, representing the balloons in a graph formalism to detect houses and the street network.

We statistically evaluated the performance of the proposed system step by step and obtained very promising results. Especially, the performance in house and street network detection in residential regions is noteworthy. These results indicate the functionality of our satellite image understanding system.

The brief summary above indicates that this book may be useful for both remote sensing and computer vision communities. For the remote sensing community, it proposes a novel end-to-end system to analyze multispectral satellite images. Hence, it may be counted as one of the pioneering works for future automated satellite and aerial image understanding systems. For the computer vision community, the book emphasizes that many new and fruitful research problems are waiting to be solved. For both communities, the book clearly shows that more collaboration between both disciplines is mandatory for developing techniques to improve human life.

Istanbul, Turkey
Troy, NY, USA

Cem Ünsalan
Kim L. Boyer

# Acknowledgements

The authors gratefully acknowledge the financial support of the US National Aeronautics and Space Administration, and meaningful discussions with Dr. Bruce Davis of the NASA Stennis Space Center, MS, in the framing and execution of this work.

# Contents

## Part IV Extracting Residential Regions



## Part V Building and Road Detection

# Chapter 1
# Introduction

Cities are evolving and districts are changing their characteristics faster than ever before. Although the evolution is slow in the central parts of most cities, it is typically fairly fast in outlying regions. Yesterday's forested or rural regions around the city become tomorrow's residential regions (These observations are especially valid in North America). These changes cause many problems for policy makers and government agencies. They affect the public and private utility networks. Maps become less reliable around these regions. As a result, emergency plans based on these maps become unreliable.

Konecny and Schiewe [1] summarize some facts regarding manual map generation. According to their analysis, 33.5% of the world was mapped at 1:25000 scale (around one meter per pixel resolution) as of 1993. This resolution is vital for mapping most mature cities (such as European cities) because their buildings and road networks are in close proximity. For this scale, the annual manual map generation rate is around 2.8%. Similarly, the annual manual map updating rate is around 4.9%. Konecny and Schiewe summarize the urgency to automate map generation:

"On average, maps of 1:25 000-scale are 20 years out of date and 1:50 000-scale sheets may be 40 (or more) years old."

Considering that a house can be built less than a year (and destroyed in minutes), and larger buildings can be built within two to three years, the inadequacy of current map updating rates is evident.

Besides affecting utility networks and maps, these changes also affect tax assessment information and demographics. The update rate of census information (10 years in the United States, for example) is insufficient to adequately track these changes.

While the advent of commercially available, high-resolution satellite imagery addresses the data collection issue, the rate at which these sensors provide data currently far exceeds the rate at which those data can be analyzed. To assist experts, planners, policy makers, and civil defense organizations, automated systems are needed.

In this book, we propose such an automated multispectral satellite image understanding system on Ikonos images. Our system has modules on land use classification, residential region detection, house and street network extraction. In developing

C. Ünsalan, K.L. Boyer, *Multispectral Satellite Image Understanding*,
Advances in Computer Vision and Pattern Recognition,
DOI 10.1007/978-0-85729-667-2_1, © Springer-Verlag London Limited 2011

automated methods for each part of the system, we extensively benefit from novel or existing computer vision and pattern recognition techniques. These techniques can also be used in future more advanced automated systems. In the following paragraphs, we provide a brief introduction to each part of our system. Then, we explore each method as well as related concepts in detail in the following chapters.

In the first part of the book, we provide detailed information on remote sensing satellites that are either operational now or that have been used extensively in the past. Since our system is based on satellite images, understanding their characteristics in the first place is mandatory. Besides, having an idea about possible satellite image sources is also necessary. This part also aims to emphasize that, more satellite images will become available in the future with higher resolution and better spectral ranges.

The second part of the book is about summarizing the multispectral information in satellite images. We focus on a set of vegetation and shadow-water indices since we use them as the source of multispectral information for land use classification and house and street network detection. Vegetation indices have been used extensively to estimate the vegetation density from satellite and airborne images for many years. We provide a thorough literature review on vegetation indices first. Then, we focus on one of the most popular vegetation indices, the normalized difference vegetation index (*NDVI*) and introduce a statistical framework to analyze it.

We propose a solution to the saturation problem of the *NDVI* based on our statistical framework. As the vegetation density increases, the corresponding *NDVI* values start to saturate and cannot represent highly vegetated regions reliably. By testing our method on real images, we show that we can obtain a linearized (and more reliable) index than the *NDVI*. We also investigate the relationship of this index with the ratio vegetation index (*RVI*), another popular vegetation index.

As we know, the *NDVI* uses only red and near-infrared bands. Using the established statistical framework, we introduce a new set of vegetation indices using blue and green bands in addition to the red and the near-infrared. We compare these indices with the measure obtained from the *NDVI* on real images and comment on the results.

The framework for the *NDVI* also serves as a basic tool for introducing a set of shadow-water indices. These indices can be extensively used to detect shadow regions in satellite images. At the same time, they can indicate water bodies in satellite images such as rivers, ponds, lakes, and the sea. In detecting house and street networks, we benefit from this index to locate lakes in residential regions.

The third part of the book focuses on methods for land use classification. The first chapter of this part summarizes and analyzes the work on land use classification from different perspectives. The reader may benefit from this information in two ways: observing the general picture in land use classification studies and inferring possible future directions for research. In the following chapters of this part, we propose novel features based on structural, multispectral, hybrid, and graph theoretical methods for land use classification. Next, we provide a brief introduction to these.

In land use classification, because the volume of data—and the size of individual images—is so great, we first focus on a "triage" in which urban and non-urban classes are first identified with (relatively) minimal processing. Therefore, we introduce a set of structural features based on straight lines to assess land development levels in high resolution (one meter) satellite images. Urban areas exhibit a preponderance of straight line features, generally appearing in fairly simple, quasiperiodic organizations. Wilderness and rural areas produce line structures in more random spatial arrangements. We use this observation to perform an initial triage on the image to restrict the attention of subsequent, more computationally intensive analyses. Statistical measures based on straight lines guide the analysis. We base these features on orientation, length, contrast, periodicity, and location. We trained and tested parametric and non-parametric classifiers using this feature set. These tests were for a two-class problem (urban vs. rural).

Although structural information can be used for land use classification alone; if multispectral information is available, it may improve the classification accuracy. Therefore, in the second chapter of part three, we benefit from vegetation indices which summarize the multispectral information in the image. We use them to improve the land use classification results. We also developed a synergistic approach that combines structural and multispectral information. In particular, the structural features serve as cue regions for multispectral features. We call these methods as hybrid to indicate that they benefit from both structural and multispectral information.

After the initial urban vs. rural classification of regions, we introduce computationally more expensive but more precise graph theoretical measures over panchromatic images especially to detect residential regions in the last chapter of part three. The graphs are constructed using straight lines (obtained from line support regions) as vertices, while graph edges encode their spatial relationships. We then introduce a set of measures based on various properties of the graph. These measures are monotonic with increasing structure (organization) in the image. Thus, increased cultural activity and land development are indicated by increases in these measures—without explicit extraction of street networks, buildings, residences, etc. These latter, time consuming tasks can be restricted only to "promising" image regions, according to our measures. We present a theoretical basis for the measures followed by extensive experimental results in which the measures are first compared to manual evaluations of land development.

In part four of the book, we extend land use classification ideas to address the problem of detecting residential regions (as the ultimate goal). To do so, some use of spatial coherence is required. Therefore, we introduce two methods. The first is based on a decision system to perform residential region classification via an overlapped voting method for consensus discovery. This method is called as feature based grouping throughout the book. The second method is based on graph theoretical measures introduced in the previous part. We then present and test a method to focus on, and (pre)extract, suburban-style residential areas. These are of particular importance in many applications, and particularly difficult to detect.

The fifth part of the book is about building and road detection. We first summarize and categorize the relevant literature on these problems. As in land use classification, this information may itself be useful to infer the future research directions for building and road network detection studies. In our multispectral satellite image understanding system, we detect houses and the street network as representative subcategories of buildings and the road network. We focused on these two subcategories since our aim is analyzing residential regions.

Our house and street network detection subsystem comprises four main parts. First, we introduce a measure on multispectral images to detect regions of possible human activity. On this measure, we introduce a variation of the k-means clustering algorithm to extract possible houses and street networks by combining both spatial and spectral features. This combination of information improves the final clustering results. From clustering, we obtain a binary image containing possible street network fragments and houses. We then decompose this binary image using a balloon algorithm based on mathematical morphology. Having obtained the decomposition, we represent the balloons in a graph theoretical framework. Balloons serve as vertices and their neighborhood information is encoded by weighted edges in the graph. The street network is extracted from the graph by using the unary and binary constraints. The remaining vertices (balloons) are assigned as possible houses in the region.

In the final part of the book, we summarize the performance of the overall multispectral satellite image understanding system. We also provide key ideas for each part. These may help the reader to improve the existing methods and introduce new automated satellite image understanding systems.

At the end of each chapter, there are review questions. These serve for two purposes. The first is to improve understanding the method(s) presented in the chapter. The second is leading the reader to more advanced methods through open ended questions. Hence, the reader will have a chance to use the proposed system as well as its novel building blocks to develop an automated satellite image understanding system for his or her needs.

## Reference

1. G. Konecny, J. Schiewe, ISPRS J. Photogramm. Remote Sens. **51**, 173 (1996)

# Part I
# Sensors

# Chapter 2
# Remote Sensing Satellites and Airborne Sensors

To develop an automated satellite image understanding system, the properties of satellite images should be known in advance. Therefore, this chapter introduces various remote sensing satellites and airborne systems. In almost all cases, the sensors on these satellites and airborne systems are called by the same name as the satellite or the airborne program. We follow the same convention.

The remote sensing satellites (sensors) we consider are Landsat, SPOT, IRS, AVHRR, Ikonos, Quickbird, FORMOSAT, CARTOSAT, Worldview, Alos, and Geoeye; the aerial sensors (programs) we consider are Daedalus, AVIRIS, HYDICE, and DAIS 7915. Wherever possible, we give a brief historical development of the sensor family with the operation dates, resolution, and revisit interval. We also give the spectral properties of the latest sensor for each family. We tabulate this information as a sensor selection guide. At the last section, we summarize and compare the properties of these sensors and their usage through time to give a brief information to the potential user.

## 2.1 Landsat

One of the best known families of remote sensing satellites is Landsat. This is a US based sensor family that has evolved over time. The first satellite in this family, launched in 1972, was Landsat 1. It had two sensors, the Return Beam Vidicon (RBV) and the Multi Spectral Scanner (MSS). RBV was a television camera, replaced by the Thematic Mapper (TM) in Landsat 4 and 5. In the last two satellites, there are panchromatic (pan), Enhanced Thematic Mapper (ETM) and Enhanced Thematic Mapper Plus (ETM+) sensors (Table 2.1). This family remains active; its average resolution is around 15 meters with a 16 day revisit interval. Table 2.2 summarizes the spectral properties of the latest family member, Landsat 7.

C. Ünsalan, K.L. Boyer, *Multispectral Satellite Image Understanding*,
Advances in Computer Vision and Pattern Recognition,
DOI 10.1007/978-0-85729-667-2_2, 

**Table 2.1** History of the Landsat family

| Satellite | Launch date | End of service | Resolution (m) | Rev. int. (days) |
|---|---|---|---|---|
| Landsat 1 | 7/23/1972 | 1/6/1978 | RBV 80; MSS 80 | 18 |
| Landsat 2 | 1/22/1975 | 2/25/1982 | RBV 80; MSS 80 | 18 |
| Landsat 3 | 3/5/1978 | 3/31/1983 | RBV 30; MSS 80 | 18 |
| Landsat 4 | 7/16/1982 | | TM 30; MSS 80 | 16 |
| Landsat 5 | 3/1/1984 | | TM 30; MSS 80 | 16 |
| Landsat 6 | 10/5/1993 | 10/5/1993 | Pan 15; ETM 30 | 16 |
| Landsat 7 | 4/15/1999 | | Pan 15; ETM+ 30 | 16 |

**Table 2.2** Landsat 7 spectral range

| Band | Spectral range (μm) | Resolution (m) |
|---|---|---|
| 1 | 0.450 to 0.515 | 30 |
| 2 | 0.525 to 0.605 | 30 |
| 3 | 0.630 to 0.690 | 30 |
| 4 | 0.750 to 0.900 | 30 |
| 5 | 1.550 to 1.750 | 30 |
| 6 | 10.400 to 12.500 | 60 |
| 7 | 2.090 to 2.350 | 30 |
| Pan | 0.520 to 0.900 | 15 |

**Table 2.3** History of the SPOT family

| Satellite | Launch date | Resolution (m) | Rev. int. (days) |
|---|---|---|---|
| SPOT 1 | 2/22/1986 | Pan 10; MS 20 | 26 |
| SPOT 2 | 1/22/1990 | Pan 10; MS 20 | 26 |
| SPOT 3 | 9/26/1993 | Pan 10; MS 20 | 26 |
| SPOT 4 | 3/4/1998 | Pan 10; MS 20 | 26 |
| SPOT 5 | 5/4/2002 | Pan 2.5 or 5; MS 10 | 26 |

## 2.2 SPOT

SPOT is a French–Belgian–Swedish joint remote sensing satellite family. SPOT 1, launched in 1986, offered 10 meter panchromatic (pan) and 20 meter multispectral (ms) images with a 26 day revisit interval. The newest member of the family has the same revisit interval with 2.5 or 5 meter panchromatic and 10 meter multispectral image resolution (Table 2.3). See Table 2.4 for the spectral properties of SPOT 5, the latest SPOT in the sky.

**Table 2.4** SPOT 5 spectral range

| Band | Spectral range (μm) | Resolution (m) |
|---|---|---|
| 1 | 0.50 to 0.59 | 10 |
| Pan | 0.48 to 0.71 | 2.5 or 5 |

**Table 2.5** The history of the IRS family

| Satellite | Launch date | End of service | Resolution (m) | Rev. int. (days) |
|---|---|---|---|---|
| IRS 1A | 3/17/1988 | 1992 | 72.5 | 22 |
| IRS 1B | 8/29/1991 | 1999 | 72.5 | 22 |
| IRS 1C | 12/28/1995 | 1997 | Pan 5.8; MS 23.5 | 24 |
| IRS 1D | 9/29/1997 | 1997 | Pan 5.8; MS 23.5 | 24 |
| IRS 2A | 2000 | | Pan 5–10; MS 23.5–70.5 | 24 |
| IRS P2 | 10/15/1994 | 1997 | Pan 5.8; MS 36.25 | 22 |
| IRS P3 | 3/21/1996 | 1997 | Pan 189; MS 523 | 5 |

**Table 2.6** IRS spectral range

| Band | Spectral range (μm) | Resolution (m) |
|---|---|---|
| 1 | 0.50 to 0.59 | 23.5 |
| 2 | 0.62 to 0.68 | 23.5 |
| 3 | 0.77 to 0.86 | 23.5 |
| 4 | 1.55 to 1.70 | 70.5 |
| Pan | 0.50 to 0.75 | 5.8 |

## 2.3 IRS

IRS is the Indian remote sensing satellite family, first launched in 1988 with 72.5 meter resolution. Although other Indian remote sensing satellites were launched prior to IRS, this was the first Indian family to see extensive use. The latest family member, IRS 2A, has 5 to 10 meter panchromatic and 23.5 to 70 meter multispectral image resolution (Table 2.5). The spectral properties of the latest family member appear in Table 2.6.

## 2.4 AVHRR

Another US based sensor is the Advanced Very High Resolution Radiometer (AVHRR). This sensor family differs from previous sensors in both resolution and intended application. It has a fairly low resolution (around 1.1 km) and is basically used for vegetation and forestry studies (Table 2.7). The latest AVHRR sensor, NOAA-16, has 6 bands covering the visible, near-infrared, and thermal infrared.

**Table 2.7** History of the AVHRR family

| Satellite | Launch date | End of service | Resolution (km) |
|---|---|---|---|
| NOAA-6 | 6/79 | 11/86 | 1.1 |
| NOAA-7 | 8/81 | 6/86 | 1.1 |
| NOAA-8 | 5/83 | 10/85 | 1.1 |
| NOAA-9 | 2/85 | 11/88 | 1.1 |
| NOAA-10 | 11/86 | 9/91 | 1.1 |
| NOAA-11 | 11/88 | 9/94 | 1.1 |
| NOAA-12 | 5/91 | 12/94 | 1.1 |
| NOAA-14 | 12/94 | | 1.1 |
| NOAA-15 | 5/98 | | 1.1 |
| NOAA-16 | 9/00 | | 1.1 |

**Table 2.8** History of the Ikonos family

| Satellite | Launch date | End of service | Resolution (m) | Rev. int. (days) |
|---|---|---|---|---|
| Ikonos 1 | 4/27/1999 | 4/27/1999 | Pan 1; MS 4 | 3 |
| Ikonos 2 | 9/24/1999 | | Pan 1; MS 4 | 3 |

**Table 2.9** Ikonos spectral range

| Band | Spectral range (μm) | Resolution (m) |
|---|---|---|
| 1 | 0.45 to 0.52 | 4 |
| 2 | 0.51 to 0.60 | 4 |
| 3 | 0.63 to 0.70 | 4 |
| 4 | 0.76 to 0.85 | 4 |
| Pan | 0.45 to 0.90 | 1 |

## 2.5 Ikonos

Ikonos is the first US based *commercial* remote sensing satellite. It has one of the highest image resolution publicly available, with one meter panchromatic and four meter multispectral images (Table 2.8). See Table 2.9 for the spectral properties of Ikonos 2.

## 2.6 Quickbird

Quickbird is another US based *commercial* remote sensing satellite. Quickbird, launched on 10/18/2001, offers 0.61 meter panchromatic (pan) and 2.44 meter multispectral (ms) images with a three day revisit interval. See Table 2.10 for the spectral properties of Quickbird.

**Table 2.10** Quickbird spectral range

| Band | Spectral range (μm) | Resolution (m) |
|---|---|---|
| 1 | 0.45 to 0.52 | 2.44 |
| 2 | 0.52 to 0.60 | 2.44 |
| 3 | 0.63 to 0.69 | 2.44 |
| 4 | 0.76 to 0.90 | 2.44 |
| Pan | 0.45 to 0.90 | 0.61 |

**Table 2.11** FORMOSAT spectral range

| Band | Spectral range (μm) | Resolution (m) |
|---|---|---|
| 1 | 0.45 to 0.52 | 8 |
| 2 | 0.52 to 0.60 | 8 |
| 3 | 0.63 to 0.69 | 8 |
| 4 | 0.76 to 0.90 | 8 |
| Pan | 0.45 to 0.90 | 2 |

## 2.7 FORMOSAT

FORMOSAT-2 is a Chinese remote sensing satellite launched on 05/21/2004. It offers two meter panchromatic (pan) and eight meter multispectral (ms) images with a one day revisit interval. See Table 2.11 for the spectral properties of FORMOSAT.

## 2.8 CARTOSAT

CARTOSAT is a recent India based remote sensing satellite launched on 05/05/2005. It only offers 2.5 meter panchromatic (pan) images with five day revisit interval.

## 2.9 Worldview

Worldview is a recent US based *commercial* remote sensing satellite family. The first member of this family offers 0.55 meter panchromatic images. The second member of this family offers 0.46 meter panchromatic and 1.8 meter multispectral images (Table 2.12). See Table 2.13 for the spectral properties of Worldview 2.

## 2.10 ALOS

ALOS is a Japanese remote sensing satellite launched on 01/24/2006. It offers 2.5 meter panchromatic (pan) and 10 meter multispectral (ms) images with a two day revisit interval. See Table 2.14 for the spectral properties of ALOS.

**Table 2.12** History of the Worldview family

| Satellite | Launch date | End of service | Resolution (m) | Rev. int. (days) |
|---|---|---|---|---|
| Worldview 1 | 9/18/2007 | | Pan 0.55 | 1.7 |
| Worldview 2 | 10/08/2009 | | Pan 0.46; MS 1.8 | 1.1 |

**Table 2.13** Worldview 2 spectral range

| Band | Spectral range (μm) | Resolution (m) |
|---|---|---|
| 1 | 0.400 to 0.450 | 1.8 |
| 2 | 0.450 to 0.510 | 1.8 |
| 3 | 0.510 to 0.580 | 1.8 |
| 4 | 0.585 to 0.625 | 1.8 |
| 5 | 0.630 to 0.690 | 1.8 |
| 6 | 0.705 to 0.745 | 1.8 |
| 7 | 0.770 to 0.895 | 1.8 |
| 8 | 0.860 to 1.040 | 1.8 |
| Pan | 0.450 to 0.800 | 0.46 |

**Table 2.14** ALOS spectral range

| Band | Spectral range (μm) | Resolution (m) |
|---|---|---|
| 1 | 0.42 to 0.50 | 10.0 |
| 2 | 0.52 to 0.60 | 10.0 |
| 3 | 0.61 to 0.69 | 10.0 |
| 4 | 0.76 to 0.89 | 10.0 |
| Pan | 0.52 to 0.77 | 2.5 |

## 2.11 Geoeye

Geoeye is yet another US based *commercial* remote sensing satellite family. The active member of this family, launched on 09/06/2010, with a three day revisit interval offers 0.41 meter panchromatic and 1.65 meter multispectral images. The second member of this family has not been launched, yet. However, it is expected to offer 0.25 meter panchromatic images. See Table 2.15 for the spectral properties of Geoeye 1.

## 2.12 Airborne Image Sensors

There are various airborne sensors available to supplement satellites for remote sensing applications. These sensors have resolutions comparable to satellite based sensors. Their superiority is the range of the spectrum they sweep and the number of

**Table 2.15** Geoeye 1 spectral range

| Band | Spectral range (μm) | Resolution (m) |
|---|---|---|
| 1 | 0.450 to 0.520 | 1.65 |
| 2 | 0.520 to 0.600 | 1.65 |
| 3 | 0.625 to 0.695 | 1.65 |
| 4 | 0.760 to 0.900 | 1.65 |
| Pan | 0.450 to 0.900 | 0.41 |

**Table 2.16** Airborne sensors

| Program | Resolution (m) | Spectral range (μm) | Total # of bands |
|---|---|---|---|
| Daedalus | 25 | 0.42 to 14.00 | 12 |
| DAIS 7915 | 3 to 20 | 0.40 to 12.60 | 79 |
| HYDICE | 1 to 4 | 0.40 to 2.45 | 210 |
| AVIRIS | 17 | 0.40 to 2.45 | 224 |

**Table 2.17** Daedalus spectral range

| Band | Spectral range (μm) | Resolution (m) |
|---|---|---|
| 1 | 0.42 to 0.45 | 25 |
| 2 | 0.45 to 0.52 | 25 |
| 3 | 0.52 to 0.60 | 25 |
| 4 | 0.60 to 0.62 | 25 |
| 5 | 0.63 to 0.69 | 25 |
| 6 | 0.69 to 0.75 | 25 |
| 7 | 0.76 to 0.90 | 25 |
| 8 | 0.91 to 1.05 | 25 |
| 9 | 1.55 to 1.75 | 25 |
| 10 | 2.08 to 2.35 | 25 |
| 11 (High gain) | 8.50 to 14.00 | 25 |
| 12 (Low gain) | 8.50 to 14.00 | 25 |

spectral bands. However, they are limited by range and time. We summarize these sensors in Table 2.16 and give the spectral properties of the Daedalus sensor in Table 2.17.

## 2.13 Summary of the Chapter

Although sensor information is probably known to most readers, we include it here for two main reasons. First, to give a review of the sensor technology. Second, most

**Table 2.18** Summary of the sensor properties

| Sensor | Resolution (m) | Spectral range (μm) | Rev. int. (days) |
|---|---|---|---|
| Landsat | 15 | 0.45 to 2.35 | 16 |
| SPOT | 2.5 | 0.50 to 1.75 | 5 |
| IRS | 5 | 0.50 to 1.70 | 5 |
| Ikonos | 1 | 0.45 to 0.85 | 3 |
| Quickbird | 0.61 | 0.45 to 0.90 | 3 |
| FORMOSAT | 2 | 0.45 to 0.90 | 1 |
| CARTOSAT | 2.5 | N/A | 5 |
| Worldview | 0.46 | 0.40 to 1.04 | 1.1 |
| ALOS | 2.5 | 0.42 to 0.89 | 2 |
| Geoeye | 0.41 | 0.45 to 0.90 | 3 |
| Airborne | 1 to 25 | 0.42 to 14.00 | N/A |

feature extraction methods directly depend on the resolution and spectrum the sensor offers. Therefore, an improvement in the sensor may lead to new methods in multispectral satellite image understanding. Let us first summarize (Table 2.18) and compare the latest family members of these sensors.

This table suggests the following conclusions. Worldview 2 and Geoeye have the highest resolution. Worldview 1 and Quickbird have the next highest resolution, followed by Ikonos and Airborne sensors. If the resolution is the most important parameter in system design, Worldview 2 seems to be the best choice available.

However, these sensors also differ in spectral range. Airborne sensors in general have the widest spectral range available. Landsat, SPOT, IRS, and Worldview 2 follow them. Although airborne sensors offer superior spectral range, they suffer from range and time (as mentioned previously). Therefore, Landsat is the optimal choice if the spectral and operational ranges are considered. However, if a higher spectral range is required, airborne sensors can be used for small geographic locations.

Worldview 2 is also superior in terms of revisit interval, followed by FORMOSAT, ALOS, Geoeye, and Ikonos. For applications requiring frequent updates, Worldview 2 offers the best solution; Landsat is the worst. Although Worldview 2 has a narrower spectral range than Landsat, its resolution and revisit interval makes this sensor attractive for most applications.

So, the potential user has a wide range of choices on the resolution, spectral range, and revisit interval he or she desires. Next, we group and count the published papers by the sensors they used in Fig. 2.1 in three year time intervals, starting from 1967 to 2002. This records the popularity of these sensors over time.

This figure indicates the popularity of the Landsat family through time, and there are many reasons for this popularity. One main reason is the high spectral range and moderate resolution Landsat offers. Another reason may be the availability of its images. Besides Landsat, airborne sensors and SPOT have also been used extensively. Since Ikonos, Quickbird, Worldview 2, and Geoeye 1 have been launched recently, they are not shown here. However, based on the above comparison these satellites

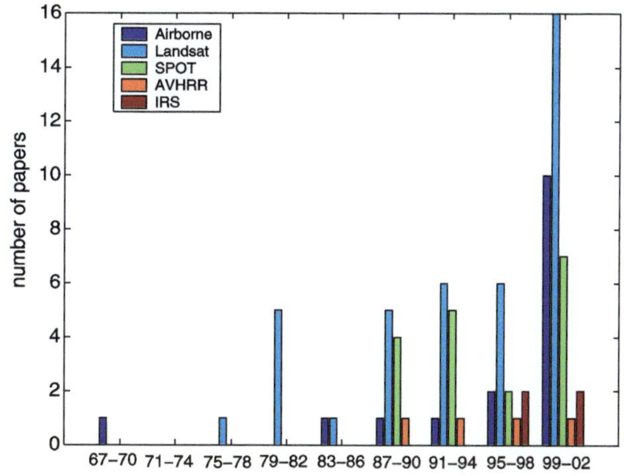


**Fig. 2.1** Sensor usage in three year time intervals from 1967 to 2002

offer great promise for future remote sensing studies. The final, somewhat obvious, observation we can make is that overall use and study of remote sensing platforms shows near-exponential growth.

## 2.14 Problems

**2.1** Summarize the satellite vs the country of origin.

**2.2** What are the differences between satellite and airborne image capturing?

**2.3** Based on the information provided in this chapter, plot

(a) The resolution of each satellite image vs its launch date.
(b) The revisit interval of each satellite vs its launch date.

**2.4** (**Open ended question**) Using the plots in the previous problem, make projections on the properties of future satellites. Comment on your projections.

# Part II
# The Multispectral Information

# Chapter 3
# Linearized Vegetation Indices

Most remote sensing satellite sensors offer multispectral images besides panchromatic images. This multispectral information is extremely valuable to detect various objects in the image. In this book, we summarize the multispectral information via vegetation and shadow–water indices. In the following chapters, we benefit from these indices in land use classification and house and street network detection. In this chapter, we start with vegetation indices.[1]

Nearly all plants need sunlight to survive, using chlorophyll to convert radiant energy from the sun into organic energy. Chlorophyll exhibits unique absorption characteristics, absorbing wavelengths around the visible red band (645 μm) while being transparent to wavelengths in the near-infrared (700 μm). These characteristics of chlorophyll are commonly used to design indices to estimate the local vegetation density in multispectral satellite or airborne imagery.

In this chapter, we consider the popular vegetation index based on the above observations, the normalized difference vegetation index (*NDVI*). Although there have been many indices proposed in the literature, the *NDVI* remains one of the most popular in remote sensing applications. Other well-known vegetation indices can be found in [2–10]. Most of these represent attempts to overcome deficiencies in the *NDVI*.

Previous studies have compared vegetation indices based on various criteria [11–19], yet no clear winner has emerged. In some experiments, the *NDVI* outperforms all other indices; for others it is less successful. It is clear that, despite many attempts to develop alternatives, the *NDVI* remains one of the most effective indicators of vegetation density.

Since its introduction, the *NDVI* has been used in a wide variety of studies including those on global vegetation [20–23], crop estimation and vegetation growth [24–28], land cover [29–38], early famine detection [39], and climate [40, 41].

[1]The figures in this chapter are obtained from our previous work [1]. Here, they appear with the kind permission of IEEE.

C. Ünsalan, K.L. Boyer, *Multispectral Satellite Image Understanding*,
Advances in Computer Vision and Pattern Recognition,
DOI 10.1007/978-0-85729-667-2_3, © Springer-Verlag London Limited 2011

Nevertheless, two main objections have been raised with respect to the *NDVI*. First, it has been argued that the *NDVI* is ad-hoc [42, 43], and our extensive literature review failed to turn up any evidence to the contrary. Second, the *NDVI* is highly nonlinear, saturating for highly vegetated areas. Many authors have mentioned this saturation problem [7, 44–48]. Although used extensively in practice, these two problems limit scientific confidence in the *NDVI*.

We address these two problems in this chapter. We first introduce a statistical framework for the *NDVI*. We then propose a solution to the nonlinearity (saturation) problem based on this statistical explanation. For completeness, we then apply the same framework to the 3D and 4D feature spaces of blue, green, red, and near-infrared bands. We work with Ikonos multispectral satellite imagery in this chapter.

## 3.1 Background and Historical Development

The pioneering work of Shull [49] examined the reflectance and absorption characteristics of leaves. Rabideau et al. [50], Billings and Morris [51], and Federer and Tanner [52] conducted experiments that indicated the same reflectance-absorption characteristics. Weidner et al. [53] conducted experiments on various types of plants (plants having thick leaves, thin leaves, dark green leaves, and desert plants) to show that these absorption characteristics are highly consistent across plant types.

Based on these observations, Jordan [54] introduced the ratio vegetation index (*RVI*) to estimate the vegetation density in a given region. Since absorption is high in the red band and low in the near-infrared, Jordan assessed vegetation density using the ratio

$$RVI = \frac{\rho_{\text{nir}}}{\rho_{\text{red}}}, \tag{3.1}$$

where $\rho_{\text{nir}}$ and $\rho_{\text{red}}$ are reflectance values in near-infrared and red bands, respectively. Colwell [55, 56] applied the *RVI* to grass canopies and found it useful for estimating biomass. One of the drawbacks of the *RVI* is that it diverges to infinity when $\rho_{\text{red}}$ goes to zero.

Nalepka et al. [57] used the square root of the *RVI*, (*SRRVI*)

$$SRRVI = \sqrt{RVI} = \sqrt{\frac{\rho_{\text{nir}}}{\rho_{\text{red}}}}. \tag{3.2}$$

This index is more linear than the *RVI*. Similarly, Rouse et al. [58] introduced the transformed vegetation index (*TVI*)

$$TVI = \sqrt{RVI + 0.5}. \tag{3.3}$$

In 1974, Rouse et al. [58] introduced the normalized difference vegetation index (*NDVI*). This index, given in (3.4), remains one of the most popular vegetation indices of all:

$$NDVI = \frac{\rho_{\text{nir}} - \rho_{\text{red}}}{\rho_{\text{nir}} + \rho_{\text{red}}}. \tag{3.4}$$

Huete [3] recognized that the *NDVI* represents a slope. He then introduced a soil adjusted vegetation index (SAVI) using this characteristic:

$$SAVI = \frac{\rho_{\text{nir}} - \rho_{\text{red}}}{\rho_{\text{nir}} + \rho_{\text{red}} + L}(1 + L), \tag{3.5}$$

where $L$ is an adjustment factor explained in detail in [3]. SAVI is another (ad hoc) attempt to address the saturation problem of the *NDVI*, which we will discuss in more detail below. Baret and Guyot [59] also observed the slope nature of the *NDVI* while studying the leaf area index.

Another vegetation index was introduced by Kauth and Thomas [2]. They introduced a transformation using four Landsat MSS bands ($d_4, d_5, d_6, d_7$). Band 5 ($d_5$) is red; bands 6 and 7 ($d_6, d_7$) are near-infrared. The Kauth–Thomas transformation is as follows:

$$\begin{bmatrix} b \\ g \\ y \\ n \end{bmatrix} = \begin{bmatrix} 0.433 & 0.632 & 0.586 & 0.264 \\ -0.290 & -0.562 & 0.600 & 0.491 \\ -0.829 & -0.522 & -0.039 & 0.194 \\ 0.223 & 0.012 & -0.543 & 0.810 \end{bmatrix} \times \begin{bmatrix} d_4 \\ d_5 \\ d_6 \\ d_7 \end{bmatrix} + \begin{bmatrix} 32 \\ 32 \\ 32 \\ 32 \end{bmatrix}. \tag{3.6}$$

In the transformed space, $b$ stands for 'brightness', $g$ stands for 'greenness', $y$ stands for 'yellowness', and $n$ stands for 'nonesuch'. Greenness is taken as a vegetation index from this transformation. In his comprehensive analysis of Landsat range and biomass monitoring, Deering [60] mentioned that the *NDVI* and the *TVI* are more significant than the vegetation index obtained from the Kauth–Thomas transformation. Jackson [61] extended the ideas of Kauth and Thomas by using n-bands. His method depends on the Gram–Schmidt orthogonalization process [62].

These transformation methods, especially that involving Gram–Schmidt orthogonalization, are the closest in philosophy to ours. However, we obtain our transformation matrices through a *statistical analysis*, which explains Huete's slope definition. We then compute (and normalize) the corresponding angle from this slope to obtain vegetation indices less susceptible to saturation. To lay the theoretical groundwork, we first review some pertinent statistical concepts.

## 3.2 Statistical Preliminaries

In this section, we briefly explain two key ideas from statistics. The first is principal components analysis (PCA), which we will use to decorrelate the random variables by an orthonormal transformation. The second is the entropy, which we will use as a basis for comparing different indices.

### *3.2.1 Principal Components Analysis (PCA)*

PCA performs a linear transformation on a set of correlated random vectors to represent them in a new space such that they are uncorrelated. The new coordinate space is related to the original by a rotation about the origin, such that the new coordinate axes are aligned with the dominant directions of dispersion (the principal components) of the data, as inferred from its covariance matrix. Our description here is necessarily terse; for more detail, see [63].

Let the original (correlated) random vectors be $\mathbf{x}_1, \mathbf{x}_2, \ldots, \mathbf{x}_n$. The linear transformation matrix $\mathbf{A}$ is calculated as follows. First, we form the data matrix $\mathbf{X}$:

$$\mathbf{X} = (\mathbf{x}_1, \mathbf{x}_2, \ldots, \mathbf{x}_n). \tag{3.7}$$

$\mathbf{X}$ is therefore a $d \times n$ matrix, where $d$ is the dimensionality of the measurement space (the number of sensor bands in use, in our case), and $n$ is the number of data vectors (observations). From this, we obtain the sample covariance matrix $\mathbf{C}_x$:

$$\mathbf{C}_x = E\big[(\mathbf{X} - \mu_{\mathbf{x}})(\mathbf{X} - \mu_{\mathbf{x}})^T\big], \tag{3.8}$$

where $\mu_{\mathbf{x}}$ is the sample mean vector, given by $\frac{1}{n}\sum_i \mathbf{x}_i$.

We then find the eigenvectors $\mathbf{e}_i$ of the sample covariance matrix, which satisfy:

$$\mathbf{C}_x \mathbf{e}_i = \lambda_i \mathbf{e}_i, \quad i = 1, 2, \ldots, d. \tag{3.9}$$

In accordance with standard practice, we index the eigenvalue–eigenvector pairs so that $\lambda_1 \geq \lambda_2 \geq \cdots \geq \lambda_d$. We also normalize the eigenvectors so that $||\mathbf{e}_i|| = 1 \; \forall \, i$. The value of $\lambda_i$ corresponds to the mean-square excursion of the data (with respect to its mean) along the direction of $\mathbf{e}_i$. The $d \times d$ transformation matrix $\mathbf{A}$ is then formed by stacking the eigenvectors, one per row:

$$\mathbf{A} = \begin{bmatrix} \mathbf{e}_1^T \\ \mathbf{e}_2^T \\ \vdots \\ \mathbf{e}_d^T \end{bmatrix}. \tag{3.10}$$

Applying this transformation to any given data vector $x_j$ is simply a matter of projecting $x_j$ onto each eigenvector:

$$\mathbf{p}_j = \mathbf{A}(\mathbf{x}_j - \mu_{\mathbf{x}}). \tag{3.11}$$

For our derivations, we shift the principal components by transformed means as follows:

$$\mathbf{pc}_j = \mathbf{A}\mathbf{x}_j, \tag{3.12}$$

where $\mathbf{pc}_j = \mathbf{p}_j + \mathbf{A}\mu_{\mathbf{x}}$; we therefore work in non-centered spaces.

### 3.2.2 Entropy

To compare different vegetation indices, we use the concept of *entropy*. The entropy measures the amount of information obtained, on average, in drawing samples from a given distribution. In effect, this captures the uncertainty, or randomness, in the experiment. Therefore, the entropy increases as the distribution becomes more uniform.

To use this property in comparison, we pick test images (such as residential region images) having high contrast between their vegetated and non-vegetated regions. Then, we obtain the approximate probability mass function (pmf) of each vegetation index for such an image. We expect a fairly uniform pmf since we selected a high contrast image. Finally, we calculate the entropy of each pmf corresponding to different vegetation indices. The higher the entropy, the better we expect that index to perform.

Let an approximate pmf be given as $h(i)$ for $i = 1, \dots, N$, $N$ being the total number of quantization levels. The corresponding entropy $E$ is:

$$E = -\sum_{i=1}^{N} \bigl[h(i) \log\bigl(h(i)\bigr)\bigr]. \tag{3.13}$$

## 3.3 Exploring the *NDVI* with a Statistical Framework

Our initial goal was to construct a new vegetation index from uncorrelated variables defined in the red–near-infrared space. We expected this new index to be more representative than the *NDVI*. However, we arrived at the *NDVI* via this construction, but gained valuable insight in the process.

### *3.3.1 Estimated PCA Transformation Matrix*

In estimating the PCA transformation matrix, we use four meter resolution multispectral Ikonos satellite imagery. We have 5905 non-overlapping images each having $200 \times 200$ pixels in blue, green, red, and near-infrared bands; we only use red and near-infrared in this section. We estimated the PCA transformation matrix $\mathbf{A}$ based on these samples to obtain:

$$\mathbf{A} = \begin{bmatrix} -0.707 & 0.707 \\ 0.707 & 0.707 \end{bmatrix}, \tag{3.14}$$

which corresponds to a $\pi/4$ radian rotation of the coordinate system. To illustrate the PCA transformation, we give the scatter plot of a sample image with eigenvectors in Fig. 3.1. In this figure, we rescaled the eigenvectors for illustration. The corresponding eigenvalues are $\lambda_1 = 1.1408$, $\lambda_2 = 0.8592$. The eigenvectors indicate the dominant (orthogonal) scatter directions for the data. Representing the data in the coordinate system defined by the eigenvectors statistically decorrelates the components. That is, the projections along $e_1$ and $e_2$ are uncorrelated random variables, while those along $\rho_{\text{red}}$ and $\rho_{\text{nir}}$ are statistically correlated.

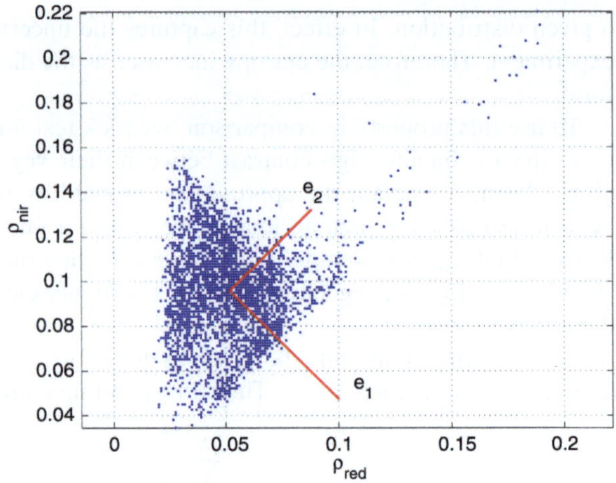


**Fig. 3.1** Scatter plot and eigenvectors for the *NDVI*

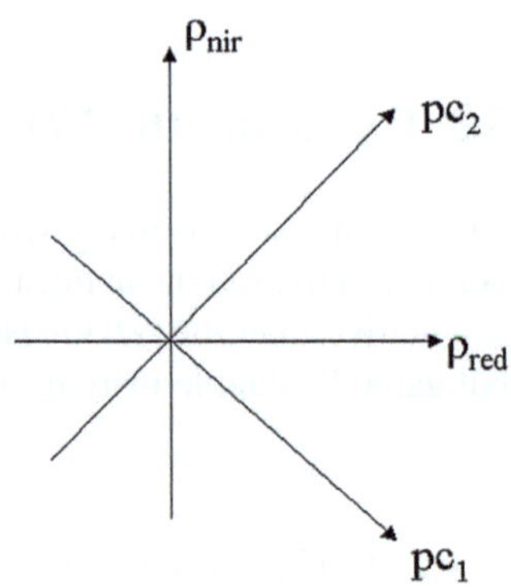


**Fig. 3.2** Rotation diagram for the PCA transformation

### *3.3.2 Statistical Construction of the* NDVI

We begin with **A**. Uncorrelated random variables $pc_1$ and $pc_2$, the projections along $e_1$ and $e_2$, respectively, are calculated as follows:

$$\begin{bmatrix} pc_1 \\ pc_2 \end{bmatrix} = \mathbf{A} \times \begin{bmatrix} \rho_{\text{red}} \\ \rho_{\text{nir}} \end{bmatrix}. \tag{3.15}$$

Figure 3.2 represents this transformation graphically. We see a $\pi/4$ radian rotation from $(\rho_{\text{red}}, \rho_{\text{nir}})$ space to $(pc_1, pc_2)$ space under this transformation.

From the construction of the PCA, we know that $pc_1$ and $pc_2$ are statistically uncorrelated. We can define a slope $\alpha$ in the new decorrelated space as:

$$\alpha = \frac{pc_1}{pc_2}. \tag{3.16}$$

If we rewrite the slope $\alpha$ in terms of $\rho_{\text{red}}$ and $\rho_{\text{nir}}$, we obtain

$$\alpha = \frac{\rho_{\text{nir}} - \rho_{\text{red}}}{\rho_{\text{nir}} + \rho_{\text{red}}}. \tag{3.17}$$

This slope definition directly corresponds to the definition of the *NDVI* given in (3.4). This derivation proves that the *NDVI* is, in fact, the slope defined with respect to the decorrelated vector space.

### *3.3.3 Saturation of the* NDVI

To this point we have (re)derived the *NDVI* by way of PCA. To our knowledge, this is the first attempt to put the *NDVI* onto a rigorous statistical footing. This derivation also suggests a solution to the saturation problem of the *NDVI*.

We begin by defining an angle $\phi$ corresponding to the slope value which is, of course, the *NDVI*:

$$\phi = \arctan(\mathit{NDVI}). \tag{3.18}$$

Next we normalize this angle by $\pi/4$ to measure vegetation density as follows:

$$\theta = \frac{4}{\pi}\phi = \frac{4}{\pi}\arctan(\mathit{NDVI}). \tag{3.19}$$

This gives a measure normalized over $\pm 1$. Equation (3.19) indicates that $\theta$ is a nonlinear transformation of the *NDVI*. By applying this nonlinear transformation, we can offset the nonlinearity of the saturation. We will show by simulation and experiments on real images how this method mitigates the saturation problem.

### *3.3.4 Experimental Results for the* NDVI *and* $\theta$

This section compares the normalized angle measure $\theta$ with the *NDVI* with respect to linearity. Ideally, one would want ground truth for comparison but, unfortunately, such data is simply unavailable to us. Therefore, we turn to indirect methods. We first evaluate the measures using a leaf area index (LAI) simulation. Although LAI is not strictly the same as vegetation density, one expects these quantities to be positively correlated under typical conditions. Then we use four real images for qualitative comparison. Finally, we quantify the comparison via three different criteria.

In future work, researchers may be able to check these methods under more-controlled conditions, should ground truth be available. Lacking this information, these criteria can assist other researchers in evaluating indices that they may develop.

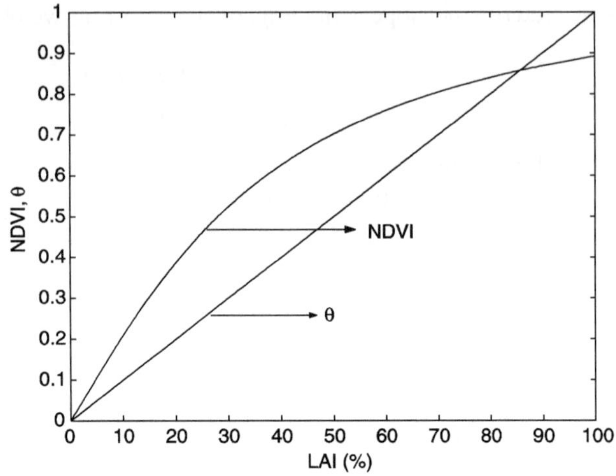


**Fig. 3.3** The *NDVI* and angle $\theta$ vs. LAI

#### 3.3.4.1 The LAI Simulation

Bégué [44] compared the *NDVI* and the LAI for various parameters and observed that it saturates. Here, we reconstruct one of the *NDVI* vs. the LAI curves of Bégué (represented in terms of percentages) and obtain $\theta$ from it. We give the reconstructed curve and $\theta$ in Fig. 3.3. Although the *NDVI* saturates as the LAI increases, $\theta$ remains highly linear.

#### 3.3.4.2 Real Images

Here, we compare the *NDVI* with the angle $\theta$ on four real images, representing various degrees of vegetation. For visual comparison, we provide color coded images (blue corresponds to the lowest vegetation value; red the highest). We also plotted the normalized histograms (pmfs) of the *NDVI* and $\theta$ for all images.

We begin with the Maryland image (Fig. 3.4). It contains almost no vegetation, except for two separate blocks (small parks). This is a typical downtown image of buildings and a street network, so the *NDVI* and $\theta$ values are low, as captured in their histograms. When the vegetation density is low, the measures respond similarly. That is, there is no consistent bias in the response of $\theta$ versus *NDVI*.

Our next example is the New Mexico image (Fig. 3.5). This image represents a typical residential region. The vegetation level is higher than that of the Maryland image, and houses and a road network are present, as well. Therefore, we expect reasonable contrast in the index value between the vegetated and non-vegetated regions of this image, suggesting an expanded dynamic range and a more uniform pmf for the better vegetation index. The figure shows that $\theta$ is, in fact, more uniformly distributed, with a wider dynamic range. The measures are highly similar in low-vegetation response, while $\theta$ shows a stronger high-vegetation response. This leads to our second conclusion: As the vegetation increases, our measure is more representative of the situation than the *NDVI*.

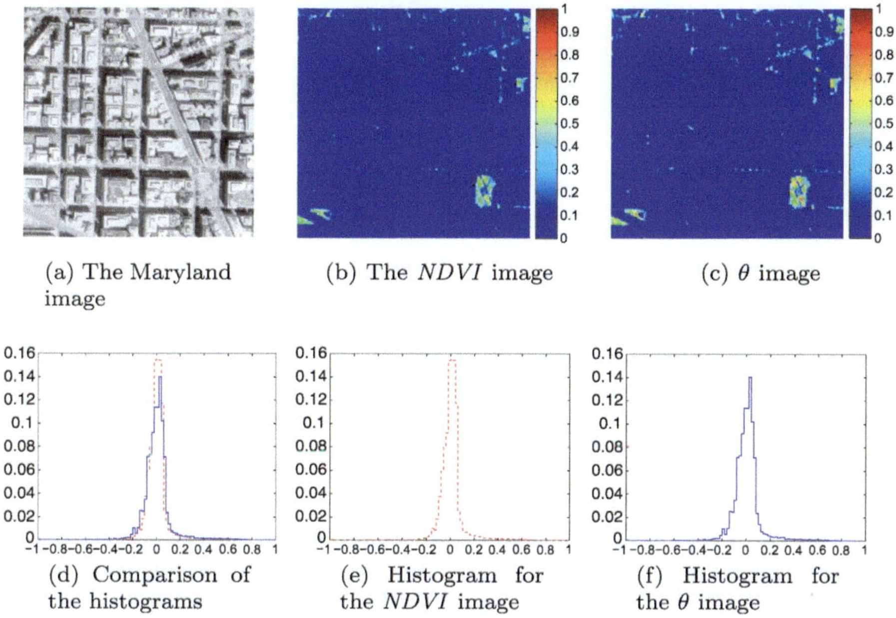


**Fig. 3.4** The Maryland image results

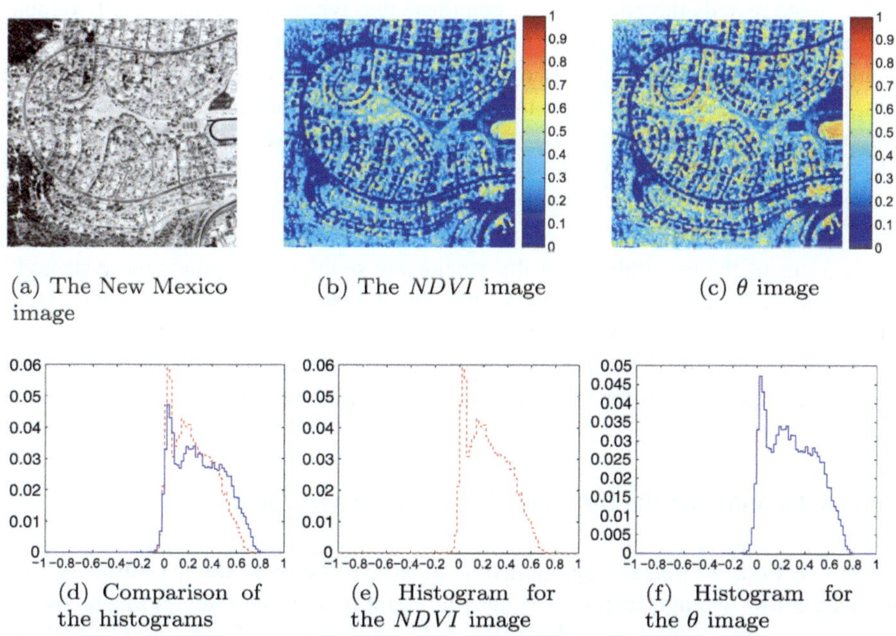


**Fig. 3.5** The New Mexico image results

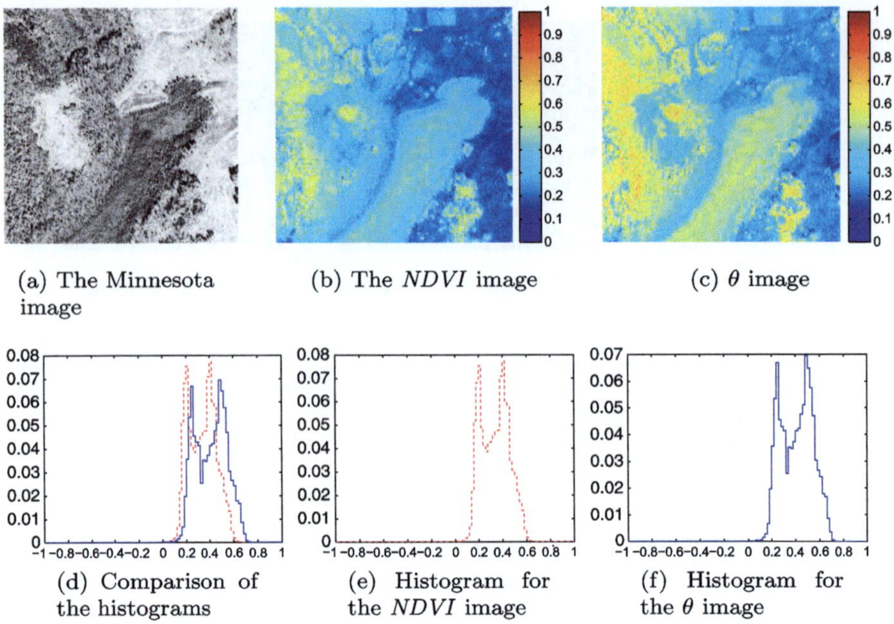


**Fig. 3.6** The Minnesota image results

Next is the Minnesota image (Fig. 3.6) containing both vegetation and bare soil. This image reveals the contrast between these two types of regions in the two measures. Both histograms are fairly bimodal, in keeping with the image content. The $\theta$ histogram again shows greater dynamic range with an extended high-vegetation response, while for bare soil the measures give similar responses. That is, in light vegetation the responses are nearly the same; in heavy vegetation they diverge as the *NDVI* saturates while $\theta$ remains highly linear.

Our last example is the Oregon image (Fig. 3.7). This is a densely forested region, one of the most heavily-vegetated in our data set. Comparing the normalized histograms of the two indices for this image, we see that $\theta$ responds more strongly to the forested region than does the *NDVI*. This indicates, again, that $\theta$ is more linear than the *NDVI*; as the *NDVI* saturates, $\theta$ continues to increase over that range.

For Figs. 3.5 and 3.6 the higher responses for $\theta$ versus *NDVI* occur only for vegetated areas; there is no consistent bias, as verified by Fig. 3.4.

#### 3.3.4.3 Quantifying the Difference Between the *NDVI* and $\theta$

In this section, we present quantitative comparisons between the measures. We anticipate a low average response to the Maryland image; the Oregon image should show high average response. The *NDVI* and $\theta$ responses are 0.0061 and 0.0072, respectively, for the Maryland image; 0.6387 and 0.7217, respectively, for the Oregon

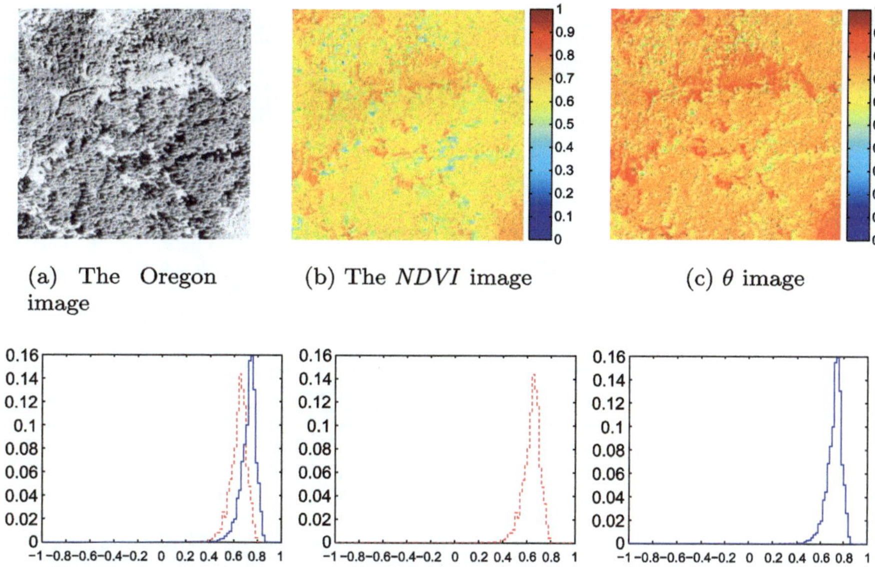


(a) The Oregon image (b) The *NDVI* image (c) $\theta$ image

(d) Comparison of the histograms (e) Histogram for the *NDVI* image (f) Histogram for the $\theta$ image

**Fig. 3.7** The Oregon image results

image. These results are in agreement with our expectations, again showing consistent behavior in low vegetation, while revealing the additional headroom available in the $\theta$ response as the vegetation increases.

Next, we introduce four new images (California 1, California 2, California 3, and Maryland 2), given in Fig. 3.8, along with the New Mexico and Minnesota images. All six of these images show high contrast between their vegetated and non-vegetated regions and cover a wide dynamic range. The entropies in the responses for these images should be indicative of index performance. The corresponding entropies are tabulated in Table 3.1. On all six images, our measure reveals higher entropy than the *NDVI*. Because both measures range over $[0, 1]$, the higher entropy indicates that the distribution of $\theta$ over the image (set) is more uniform than that of the *NDVI*. That is, $\theta$ makes better use of its available numerical range. In large part, this is because *NDVI* reaches unity only when $\rho_{\text{red}} = 0$ and $\rho_{\text{nir}} \neq 0$, which is an extremely rare occurrence, regardless of vegetation level. Therefore, the entropy test underscores the superior linearity of our measure with respect to the *NDVI*.

## 3.4 Using the Statistical Framework to Develop New Indices

Having such a powerful statistical framework to analyze the *NDVI* suggests that more vegetation indices (or indices for other ground covers) may be defined via the same process. The framework offers a straightforward mechanism for exploring

**Fig. 3.8** Additional images for entropy comparison

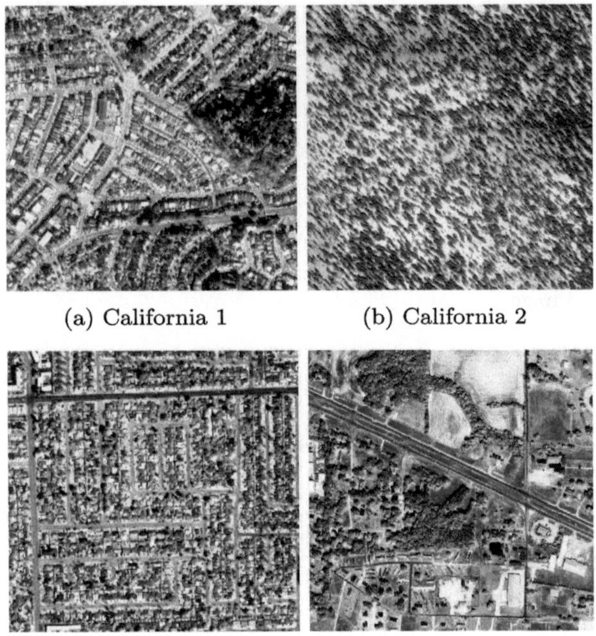

(a) California 1 (b) California 2

(c) California 3 (d) Maryland 2

**Table 3.1** Entropy comparison (bits)

| | $E_{NDVI}$ | $E_\theta$ |
|---|---|---|
| California 1 | 4.9145 | 5.1858 |
| California 2 | 4.8242 | 5.0849 |
| New Mexico | 4.9825 | 5.2148 |
| Minnesota | 4.4330 | 4.6091 |
| California 3 | 4.9156 | 5.1755 |
| Maryland 2 | 5.1361 | 5.2844 |
| Average | 4.8677 | 5.0924 |

index possibilities in spaces of higher dimensionality. The *NDVI* uses only the red and near-infrared bands. To these, we now add the blue and green bands to discover whether we can obtain more (or better) information from them.

We start by adding the blue band to the red and near-infrared to compute a transformation in this 3D space. We then replace the blue band with the green and repeat the process. Finally, we consider all four bands to determine whether a better vegetation index results in the 4D space. All measures are normalized between $\pm 1$ as is $\theta$. In this section, we compare each new index with $\theta$ on six images.

We cannot offer a physical explanation for the new indices introduced in this section at present. Our main focus is on the statistical framework as a means of "index discovery". We believe that this can provide another perspective to the remote

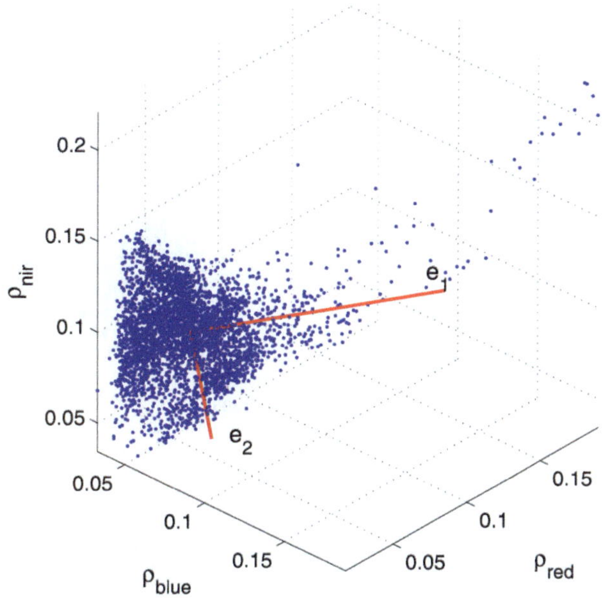


**Fig. 3.9** Scatter plot and eigenvectors for the blue, red, and near-infrared bands; the eigenvectors are rescaled for illustration; the third eigenvector is obscured

sensing community in developing vegetation (or other) indices. Of course, physical explanations are desirable (perhaps even preferred), but are not (necessarily) essential for the discovery of effective indices.

## *3.4.1 Using the Blue, Red, and Near-Infrared Bands*

To reduce atmospheric effects, many authors include the blue band in their vegetation indices. To see how this could be used in our approach, we introduced the blue band alongside the red and near-infrared bands. That is, we estimated a new PCA transformation matrix, and applied this to the three dimensional space as we did for the *NDVI* derivation to arrive at the transformed space:

$$\begin{bmatrix} pc_{21} \\ pc_{22} \\ pc_{23} \end{bmatrix} = \begin{bmatrix} -0.596 & -0.603 & -0.530 \\ 0.417 & 0.332 & -0.846 \\ 0.686 & -0.725 & 0.054 \end{bmatrix} \times \begin{bmatrix} \rho_{\text{blue}} \\ \rho_{\text{red}} \\ \rho_{\text{nir}} \end{bmatrix}. \tag{3.20}$$

Figure 3.9 gives the scatter plot of a sample image in blue–red–near-infrared space with eigenvectors. The eigenvectors are rescaled for illustration. The corresponding eigenvalues are $\lambda_1 = 1.9734$, $\lambda_2 = 0.9741$, and $\lambda_3 = 0.0525$. Since the first two eigenvalues are much greater than the third, the data scatter primarily over a 2D manifold in $\Re^3$. Thus, even in 3D, this is essentially a 2D problem. However, the 2D space in question is not spanned by $\rho_{\text{red}}$ and $\rho_{\text{nir}}$; it is spanned by two linear combinations of $\rho_{\text{blue}}$, $\rho_{\text{red}}$ and $\rho_{\text{nir}}$. Inspired by the *NDVI* (and experimentally

**Fig. 3.10** Scatter plot and eigenvectors for the green, red, and near-infrared bands; the eigenvectors are rescaled for illustration; the third eigenvector is obscured

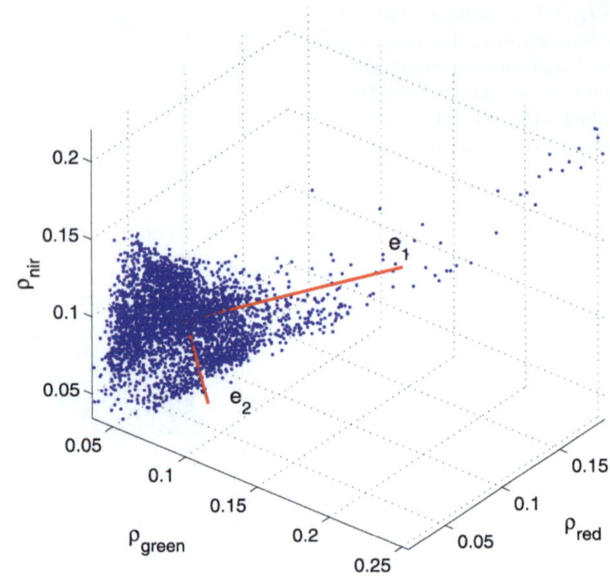


checking its performance compared to other combinations), we choose the angle between the second eigenvector and the sum vector as a new vegetation index in this space:

$$\theta_2 = \frac{4}{\pi} \arctan\left( \frac{-pc_{22}}{\sqrt{pc_{21}^2 + pc_{22}^2 + pc_{23}^2}} \right). \tag{3.21}$$

This index is similar to the $\theta$ formulation we obtained from the *NDVI*. The ratio inside the arctangent is again a slope in the transformed space. By using the angle representation instead of this slope, we again linearize the resulting index.

### 3.4.2 Using the Green, Red, and Near-Infrared Bands

Following the same procedure, we obtain the transformed space:

$$\begin{bmatrix} pc_{31} \\ pc_{32} \\ pc_{33} \end{bmatrix} = \begin{bmatrix} -0.599 & -0.600 & -0.531 \\ 0.384 & 0.367 & -0.847 \\ 0.703 & -0.711 & -0.011 \end{bmatrix} \times \begin{bmatrix} \rho_{green} \\ \rho_{red} \\ \rho_{nir} \end{bmatrix}. \tag{3.22}$$

Figure 3.10 presents the scatter plot of a sample image with (rescaled) eigenvectors in this space. The corresponding eigenvalues are $\lambda_1 = 2.0226$, $\lambda_2 = 0.9511$, and $\lambda_3 = 0.0263$; similar observations regarding the scatter (2D manifold in $\Re^3$) apply. In this space, the difference between the first two eigenvalues is greater. The last eigenvalue is also smaller than for the previous space. We again choose the angle between the second eigenvector and the sum vector as a new index in this space:

$$\theta_3 = \frac{4}{\pi}\arctan\left(\frac{-pc_{32}}{\sqrt{pc_{31}^2 + pc_{32}^2 + pc_{33}^2}}\right). \tag{3.23}$$

This index exhibits characteristics similar to those of $\theta_2$.

### 3.4.3 Using All Four Bands

Finally, we use all four bands to obtain yet another vegetation index. In doing so, we hope to tap all the information these multispectral images can provide. Applying the statistical framework to this space, we obtain:

$$\begin{bmatrix} pc_{41} \\ pc_{42} \\ pc_{43} \\ pc_{44} \end{bmatrix} = \begin{bmatrix} -0.519 & -0.526 & -0.525 & -0.422 \\ 0.309 & 0.211 & 0.209 & -0.903 \\ 0.703 & -0.043 & -0.706 & 0.067 \\ 0.375 & -0.823 & 0.426 & 0.035 \end{bmatrix} \times \begin{bmatrix} \rho_{\text{blue}} \\ \rho_{\text{green}} \\ \rho_{\text{red}} \\ \rho_{\text{nir}} \end{bmatrix}. \tag{3.24}$$

The eigenvalues are $\lambda_1 = 2.9636$, $\lambda_2 = 0.9756$, $\lambda_3 = 0.0538$, and $\lambda_3 = 0.0070$. There are again two dominant principal components, with fairly large eigenvalues. The difference between the first and the second eigenvalue has increased compared to the three dimensional transformations. The third eigenvalue is similar in value to those of the three dimensional transformations. The fourth eigenvector is fairly insignificant. This is still a 2D problem, even in a 4D space. For the vegetation index, we reason as before to get

$$\theta_4 = \frac{4}{\pi}\arctan\left(\frac{-pc_{42}}{\sqrt{pc_{41}^2 + pc_{42}^2 + pc_{43}^2 + pc_{44}^2}}\right). \tag{3.25}$$

This index also exhibits characteristics similar to those of $\theta_2$ and $\theta_3$.

## 3.5 Comparing the Vegetation Indices

To this point, we have introduced a linearized vegetation index $\theta$ based on the *NDVI*, followed by three new vegetation indices: $\theta_2$ and $\theta_3$ defined in 3D spaces, and $\theta_4$ defined in the full 4D space. Here, we compare these four indices using three different criteria. The first criterion is the dynamic range, assessed both qualitatively (visual comparison of pseudocolor images) and through normalized response histograms. The second criterion considers the entropy for the six high contrast images given in Sect. 3.3.4. The third criterion is the computational cost.

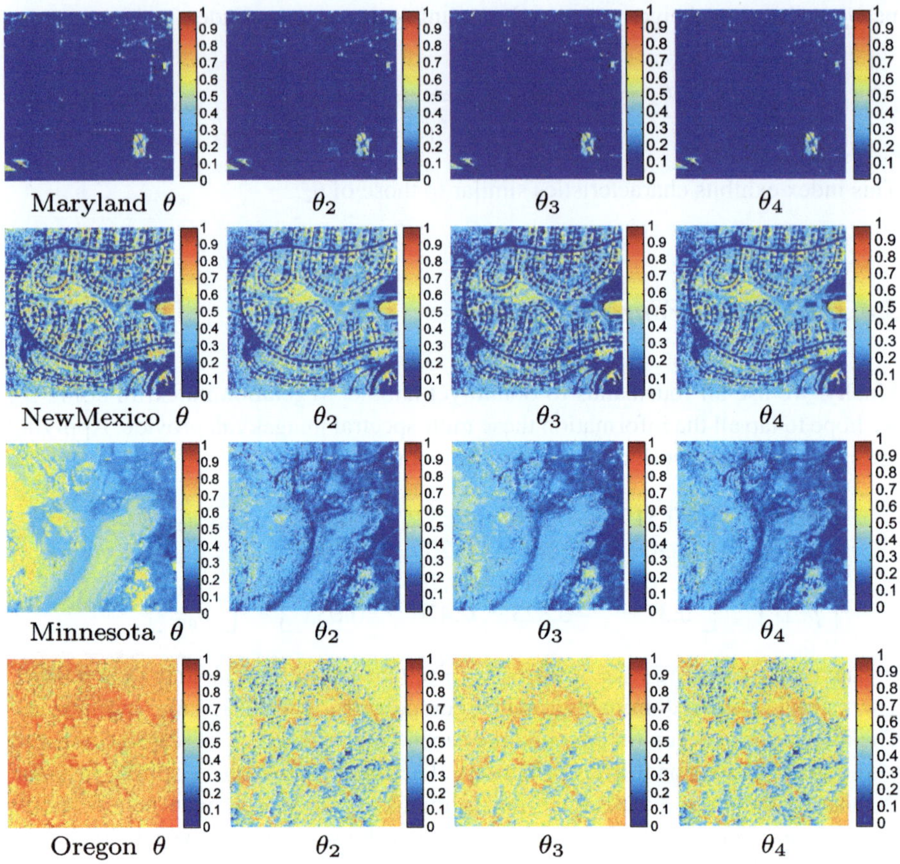


**Fig. 3.11** Comparison of vegetation indices on real images

### *3.5.1 Visual Comparison and Dynamic Range*

This section compares the indices on the Maryland, New Mexico, Minnesota, and Oregon images (Figs. 3.4, 3.5, 3.6, 3.7). As mentioned above, these images were selected and sorted based on the relative degree of vegetation. We give the color-coded responses (blue corresponds to the lowest vegetation value, red corresponds the highest) of all four vegetation indices in Fig. 3.11. In the colormap, negative index values (caused by some vegetation-free surfaces such as ice, rock, or concrete) are all mapped to 0. This reduces the visual dynamic range and makes some of the differences among the indices difficult to discern by visual inspection alone. Nevertheless, some trends can be observed.

Figure 3.11 shows that all four indices respond reasonably to vegetation. Those indices not using the blue band, $\theta$ and $\theta_3$, do not respond as well in the shadowed areas of the Maryland and Oregon images. That is, vegetation in shadow confuses these indices to some degree. The other two, $\theta_2$ and $\theta_4$, suggest the presence of

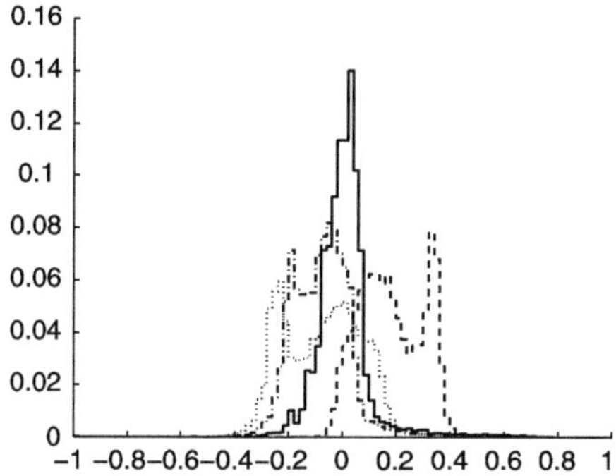


(a) Maryland

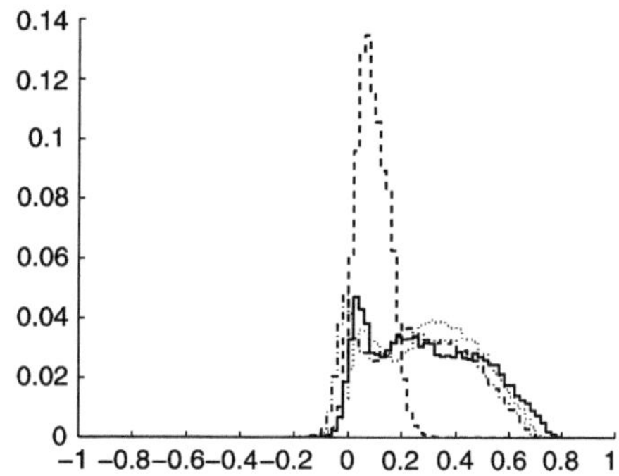


(b) New Mexico

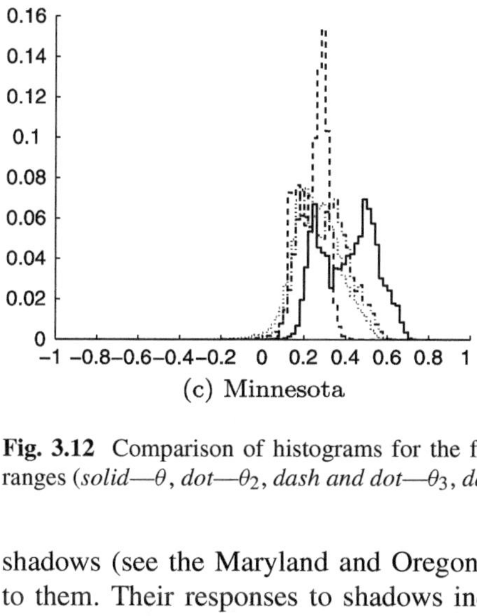


(c) Minnesota

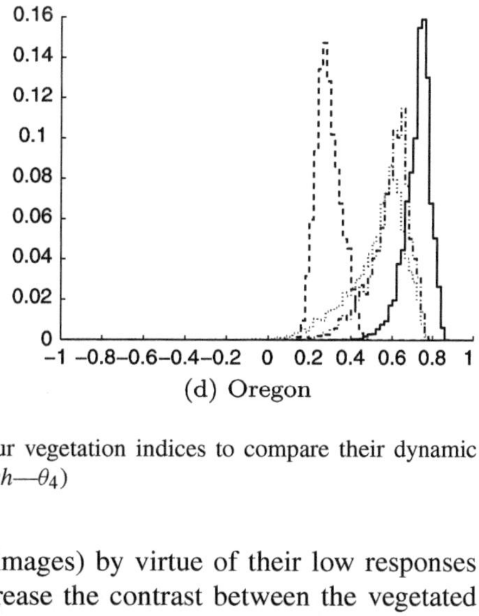


(d) Oregon

**Fig. 3.12** Comparison of histograms for the four vegetation indices to compare their dynamic ranges (*solid*—$\theta$, *dot*—$\theta_2$, *dash and dot*—$\theta_3$, *dash*—$\theta_4$)

shadows (see the Maryland and Oregon images) by virtue of their low responses to them. Their responses to shadows increase the contrast between the vegetated and non-vegetated (especially shadowed) regions. (This is not easily seen in the pseudocolors because of the truncation of negative responses.) So it appears that an index having the blue band in its formulation ($\theta_2$ or $\theta_4$) may offer some advantage in dealing with shadows. As we will see below, $\theta_4$ will prove unsuitable from a dynamic range standpoint.

To compare the dynamic range of these four vegetation indices, we consider the normalized histograms. Figure 3.12 gives the histograms for the four images; the solid curve represents $\theta$, the dotted curve $\theta_2$, the dashed and dotted curve $\theta_3$, and the dashed curve $\theta_4$.

In comparing these histograms, we see that $\theta_4$ has the lowest dynamic range, except for the nearly vegetation-free Maryland image. In the two high contrast images (New Mexico, and Minnesota) $\theta$ shows the greatest dynamic range. For these images, $\theta_2$ and $\theta_3$ give similar responses. For the Oregon image, $\theta$ has a lower dynamic range, but this is a low-contrast image of heavily-vegetated terrain. The range

**Table 3.2** Comparison of the four vegetation indices on the Maryland and Oregon images

| | $\mu_\theta$ | $\mu_{\theta_2}$ | $\mu_{\theta_3}$ | $\mu_{\theta_4}$ |
|---|---|---|---|---|
| Maryland | 0.0072 | −0.0658 | −0.0743 | 0.1835 |
| Oregon | 0.7217 | 0.5286 | 0.5861 | 0.2903 |

**Table 3.3** Comparison of the entropies on six images for the four vegetation indices

| | $E_\theta$ | $E_{\theta_2}$ | $E_{\theta_3}$ | $E_{\theta_4}$ |
|---|---|---|---|---|
| California 1 | 5.1858 | 5.2388 | 5.1323 | 4.0726 |
| California 2 | 5.0849 | 4.9506 | 4.9629 | 3.0714 |
| New Mexico | 5.2148 | 5.1381 | 5.1487 | 3.5988 |
| Minnesota | 4.6091 | 4.4696 | 4.4801 | 3.6012 |
| California 3 | 5.1755 | 5.2649 | 5.1696 | 3.6369 |
| Maryland 2 | 5.2844 | 5.0725 | 5.2083 | 3.9555 |
| Average | 5.0924 | 5.0224 | 5.0170 | 3.6561 |

of $\theta$ response close to one, as it should be. Again, for this image $\theta_2$ and $\theta_3$ respond similarly. $\theta_4$ performed poorly, giving the weakest response on this image.

Next, we focus on the two extremes (Maryland and Oregon) to compare the responses of the four vegetation indices by their sample means, $\mu$. Because the Maryland image has almost no vegetation, the proper response to this image should be near zero. The Oregon image has the greatest vegetation density, so a strong response is expected. We give these comparisons in Table 3.2.

According to the results in Table 3.2, we sort the indices from best to worst as: $\theta$, $\theta_3$, $\theta_2$, and $\theta_4$. In particular, even the dramatic difference in vegetation content between the two images produces little difference in the response of $\theta_4$, while (somewhat) diminished sensitivity is observed for $\theta_3$ and $\theta_2$.

### *3.5.2 Comparison by the Entropy on High Contrast Images*

Next, we compare the four indices based on the entropy of the response for the six high contrast images of Sect. 3.3.4 in Table 3.3. This table indicates that $\theta$ has the highest average entropy over these six images. The entropy for $\theta_2$ and $\theta_3$ are very similar, in the mid-range. Finally, $\theta_4$ is noteworthy in having significantly lower entropy. Based on this criterion, we sort these indices: $\theta$, $\theta_2$, $\theta_3$, and $\theta_4$, from best to worst.

### *3.5.3 Computational Cost*

There is a computational cost associated with each index. Our first measure, $\theta$, needs only the red and near-infrared bands. Indices $\theta_2$ and $\theta_3$ need three bands, while $\theta_4$

requires four. As we mentioned, if atmospheric effects are considered $\theta_2$ may be the best measure since it uses the blue band and performs about as well as $\theta$ in other conditions.

To summarize, $\theta$ and $\theta_2$ are the most promising two vegetation indices considered in this chapter. The ultimate selection of a "best" measure will depend to some extent on the application. To recap, we give these two most promising indices below:

$$\theta = \frac{4}{\pi}\arctan\left(\frac{\rho_{\text{nir}} - \rho_{\text{red}}}{\rho_{\text{nir}} + \rho_{\text{red}}}\right), \tag{3.26}$$

$$\theta_2 = \frac{4}{\pi}\arctan\left(\frac{-0.4167\,\rho_{\text{blue}} - 0.3317\,\rho_{\text{red}} + 0.8464\,\rho_{\text{nir}}}{\sqrt{\rho_{\text{blue}}^2 + \rho_{\text{red}}^2 + \rho_{\text{nir}}^2}}\right). \tag{3.27}$$

## 3.6 Summary of the Chapter

In this chapter, we first introduced a statistical framework for the *NDVI*, to provide it with a rigorous theoretical basis. Although the original definition seems to have been somewhat ad hoc, we were able to provide a sound theoretical justification for the index. To our knowledge, there have been no prior studies establishing a statistical foundation for the *NDVI*.

We then propose a solution to the nonlinearity (saturation) problem based on this statistical development. By representing the *NDVI* as a slope and using its inverse tangent, we could linearize the measure to yield a new index. We presented simulations and real image results to substantiate our claim.

For completeness, we applied our statistical framework to 3D and 4D feature spaces of blue, green, red, and near-infrared bands. The 2D feature space used in developing the *NDVI* was based on physical arguments alone. We developed vegetation indices based on slope definitions in the 3D and 4D transformed spaces similar to the *NDVI* although these new indices were not as powerful as the linearized *NDVI*. However, with these new indices we can use information in the blue band to mitigate the atmospheric effects observed by many authors. We also point out that our investigation of candidate slopes in the 3D and 4D transformed spaces is not exhaustive; other slope definitions may yet be found in these spaces that outperform $\theta_2$, $\theta_3$, and $\theta_4$, respectively.

If we check other popular vegetation indices in the literature, we observe that most of them are based on ratios, as is the *NDVI*. These indices can also be viewed as slopes in transformed spaces. Therefore, the transformations corresponding to these vegetation indices are possible candidates to be explored by our statistical framework.

## 3.7 Problems

**3.1** Can *RVI*, *TVI*, $SRVI$ be written in terms of *NDVI*?

**3.2** What is the most popular vegetation index in the literature? Why?

**3.3** Does PCA depend on a linear or a nonlinear transformation?

**3.4** Why is saturation and linearity of a vegetation index important?

**3.5** Which linearized vegetation index gives the best result?

**3.6** (**Open ended question**) Instead of using PCA in the linearized vegetation index derivation, does using independent components analysis (ICA), or another nonlinear transformation, lead to a stronger vegetation index?

## References

1. C. Ünsalan, K.L. Boyer, IEEE Trans. Geosci. Remote Sens. **42**(4), 907 (2004)
2. R.J. Kauth, G.S. Thomas, in *Proceedings of the Symposium Machine Processing of Remote Sensing Data, LARS, Purdue* (1976)
3. A.R. Huete, Remote Sens. Environ. **25**, 295 (1988)
4. D. Tanré, B.N. Holben, Y.J. Kaufman, IEEE Trans. Geosci. Remote Sens. **30**(2), 231 (1992)
5. Y.J. Kaufman, D. Tanré, IEEE Trans. Geosci. Remote Sens. **30**(2), 261 (1992)
6. H.Q. Liu, A.R. Huete, IEEE Trans. Geosci. Remote Sens. **33**(2), 457 (1995)
7. A.R. Huete, H.Q. Liu, K. Batcily, W.V. Leeuwen, Remote Sens. Environ. **59**, 440 (1997)
8. A.R. Huete, C. Justice, W.V. Leeuwen, MODIS Vegetation Index (MOD13), EOS MODIS Algorithm. Theoretical basis document, NASA Goddard Space Flight Center, Greenbelt, MD (1996)
9. B. Pinty, M.M. Verstraete, Vegetatio **101**, 15 (1992)
10. M.M. Verstraete, B. Pinty, IEEE Trans. Geosci. Remote Sens. **34**(5), 1254 (1996)
11. G. Rondeaux, M. Steven, F. Baret, Remote Sens. Environ. **55**, 95 (1996)
12. K. McGwire, T. Minor, L. Fenstermaker, Remote Sens. Environ. **72**, 360 (2000)
13. A.J. Elmore, J.F. Mustard, S.J. Manning, D.B. Lobell, Remote Sens. Environ. **73**, 87 (2000)
14. C.D. Elvidge, Z. Chen, Remote Sens. Environ. **54**, 38 (1995)
15. A.R. Huete, H.Q. Liu, IEEE Trans. Geosci. Remote Sens. **32**(4), 897 (1994)
16. V. Jayaraman, S.K. Srivastava, D.K. Raju, U.R. Rao, IEEE Trans. Geosci. Remote Sens. **38**(1), 587 (2000)
17. R.L. Lawrence, W.J. Ripple, Remote Sens. Environ. **64**, 91 (1998)
18. M. Roderick, R. Smith, S. Cridland, Remote Sens. Environ. **56**, 57 (1996)
19. N.H. Broge, E. Leblanc, Remote Sens. Environ. **76**, 156 (2000)
20. S.N. Goward, C.J. Tucker, D.G. Dye, Vegetatio **64**, 3 (1985)
21. C.J. Tucker, P.J. Sellers, Int. J. Remote Sens. **7**, 1395 (1986)
22. J.P. Malingreau, C.J. Tucker, N. Laporte, Int. J. Remote Sens. **10**, 855 (1989)
23. J.R.G. Townshend, C.O. Justice, D. Skole, Int. J. Remote Sens. **15**, 3417 (1994)
24. R.B. Myneni, G. Asrar, D. Tanré, B.J. Choudhury, IEEE Trans. Geosci. Remote Sens. **30**(2), 302 (1992)
25. P.S. Thenkabail, R.B. Smith, E. De Pauw, Remote Sens. Environ. **71**, 158 (2000)
26. G.B. Senay, R.L. Elliot, For. Ecol. Manag. **128**, 83 (2000)
27. T.N. Carlson, D.A. Ripley, Remote Sens. Environ. **62**, 241 (1997)
28. C.O. Justice, J.R.G. Townshend, B.N. Holben, C.J. Tucker, Int. J. Remote Sens. **6**, 1271 (1985)
29. J.R.G. Townshend, C. Justice, W. Li, C. Gurney, J. McManus, Remote Sens. Environ. **35**, 243 (1991)

30. V. Caselles, E. Valor, Remote Sens. Environ. **57**, 167 (1996)
31. J. Symnazik, R.R. Griffiths, L. Gillies, in *Proceedings of the Statistical Computing Section and Section on Statistical Graphics* (2000), pp. 10–19
32. Y.J. Kaufman, L.A. Remer, IEEE Trans. Geosci. Remote Sens. **32**(3), 672 (1994)
33. T. Fung, in *Proceedings of IGARSS*, vol. 2 (1997), pp. 836–838
34. J.M.C. Pereira, IEEE Trans. Geosci. Remote Sens. **37**(1), 217 (1999)
35. P.R. Coppin, M.E. Bauer, IEEE Trans. Geosci. Remote Sens. **32**(4), 918 (1994)
36. D.F. Lozano-Garcia, R.N. Fernández, C.J. Johannsen, IEEE Trans. Geosci. Remote Sens. **29**(2), 331 (1991)
37. T. Ishiyama, S. Tanaka, K. Uchida, S. Fujikawa, Y. Yamsahita, M. Kato, Adv. Space Res. **28**(1), 183 (2001)
38. P.M. Teillet, K. Staenz, D.J. Willimas, Remote Sens. Environ. **61**, 139 (1997)
39. F.N. Kogan, Adv. Space Res. **15**(11), 91 (1995)
40. P.J. Sellers, C.J. Tucker, G.J. Collatz, Int. J. Remote Sens. **15**, 3519 (1994)
41. Z. Li, M. Kafatos, Remote Sens. Environ. **71**, 239 (2000)
42. J.C. Price, IEEE Trans. Geosci. Remote Sens. **31**(3), 727 (1993)
43. W.C. Bausch, J.C. Price, Remote Sens. Environ. **52**, 55 (1995)
44. A. Bégué, Remote Sens. Environ. **46**, 45 (1993)
45. D.G. Brown, in *GIS and Remote Sensing Applications in Biogeography and Ecology*, ed. by A. Millington, S.J. Walsh, P. Osborne (Kluwer, Dordrecht, 2001), pp. 7–22
46. C.L. Wiegand, S.J. Mass, J.K. Aase, J.L. Hatfield, P.J.J. Pinter, R.D. Jackson, E.T. Kanemasu, R.L. Lapitan, Remote Sens. Environ. **42**, 1 (1992)
47. T. Miura, A.R. Huete, H. Yoshioka, IEEE Trans. Geosci. Remote Sens. **38**(3), 1399 (2000)
48. R.K. Gupta, T.S. Prasad, D. Vijayan, Adv. Space Res. **26**(7), 1047 (2000)
49. C.A. Shull, Bot. Gaz. **87**(5), 583 (1929)
50. G.S. Rabideau, C.S. French, A.S. Holt, Am. J. Bot. **33**(10), 769 (1946)
51. W.D. Billings, R.J. Morris, Am. J. Bot. **38**(5), 327 (1951)
52. C.A. Federer, C.B. Tanner, Ecology **47**(4), 555 (1966)
53. D.M. Gates, J.C.S. Keegan, V.R. Weidner, Appl. Opt. **4**(1), 11 (1965)
54. C.F. Jordan, Ecology **50**(4), 663 (1969)
55. J.E. Colwell, Remote Sens. Environ. **3**, 175 (1974)
56. J.E. Colwell, Bidirectional spectral reflectance of grass canopies for determination of above ground standing biomass. PhD thesis, University of Michigan (1973)
57. R.F. Nalepka, C.E. Colwell, D.P. Rice, Forecasts of winter wheat yield and production using LANDSAT data. Final Report 114800-38-F, NASA CR/ERIM (1977)
58. J.W. Rouse, R.H. Haas, J.A. Schell, D.W. Deering, J.C. Harlan, Monitoring the vernal advancement of natural vegetation. Final report, NASA/GCSFC, Greenbelt, MD (1974)
59. F. Baret, G. Guyot, Remote Sens. Environ. **35**, 161 (1991)
60. D.W. Deering, Rangeland reflectance characteristics measured by aircraft and spacecraft sensors. PhD thesis, Texas AM University, College Station, Texas (1978)
61. R.D. Jackson, Remote Sens. Environ. **13**, 409 (1983)
62. G. Strang, *Linear Algebra and Its Applications*, 3rd edn. (Harcourt College Publishers, Florida, 1986)
63. I.T. Joliffe, *Principal Component Analysis*, 2nd edn. (Springer, New York, 2002)

# Chapter 4
# Linearized Shadow and Water Indices

The next set of indices we introduce within the same statistical framework used in the previous chapter are the shadow–water indices (SWI). We benefit from these indices to detect lakes in residential regions in the following chapters. Therefore, they also provide valuable information in analyzing multispectral images.

We were unable to find a focused work on water detection in satellite imagery. However, the closest matches in the literature are [1–3]. Shadows (clouds) also have photometric characteristics similar to water. Shadow (cloud) detection studies include [4–10]. Simpson and Stitt [7] worked on cloud shadow detection in AVHRR imagery. To detect shadows, they used geometric and optical constraints on a pixel basis in multispectral images. They discarded water regions to focus on clouds.

In the Ikonos spectrum, water shows an increasing response curve until the blue band, it reaches the maximum in this region and then decreases monotonically to the near-infrared [11]. So a representative shadow–water index should be composed of high blue values first. Ideally, it should also consider the green and red bands, but the green band also responds strongly to vegetation and this impairs the shadow or water observation. Hence, the index should include blue and red bands at least. To obtain such an index, we applied the same framework we used for the *NDVI* derivation using principal components analysis with the blue, red, and near-infrared bands. Based on the combinatorial search (and trying to maximize blue and red band coefficients) we obtain the best performing shadow–water index for each dimension.

We start with a two-dimensional space. Since we plan to have the blue band in a candidate SWI, we replace the red band with the blue in the *NDVI* formulation. We obtain our first index as:

$$\gamma_1 = \frac{4}{\pi} \arctan\left( \frac{\rho_{\mathrm{nir}} - \rho_{\mathrm{blue}}}{\rho_{\mathrm{nir}} + \rho_{\mathrm{blue}}} \right). \tag{4.1}$$

For the three- and four-dimensional spaces, we observe that the third principal component has the shadow information in all the transformed spaces. Shadows and water are relatively uncommon in our training data, so the mean-square energy in that component will be less than of two—but not least of four—components. We give the corresponding shadow–water indices $\gamma_2$, $\gamma_3$, and $\gamma_4$ obtained from the transformation spaces given in (3.20), (3.22), and (3.24) as:

C. Ünsalan, K.L. Boyer, *Multispectral Satellite Image Understanding*,
Advances in Computer Vision and Pattern Recognition,
DOI 10.1007/978-0-85729-667-2_4, 

$$\gamma_2 = \frac{4}{\pi}\arctan\left(\frac{-pc_{23}}{\sqrt{pc_{21}^2 + pc_{22}^2 + pc_{23}^2}}\right), \tag{4.2}$$

$$\gamma_3 = \frac{4}{\pi}\arctan\left(\frac{-pc_{33}}{\sqrt{pc_{31}^2 + pc_{32}^2 + pc_{33}^2}}\right), \tag{4.3}$$

$$\gamma_4 = \frac{4}{\pi}\arctan\left(\frac{-pc_{43}}{\sqrt{pc_{41}^2 + pc_{42}^2 + pc_{43}^2 + pc_{44}^2}}\right). \tag{4.4}$$

## 4.1 Comparing the Shadow-Water Indices

Similar to Sect. 3.5, we compare the four SWI ($\gamma_1$, $\gamma_2$, $\gamma_3$, and $\gamma_4$) in this section. Here, we only use the first two criteria, since computation costs for the SWI are the same as for the corresponding vegetation indices.

### *4.1.1 Comparison by the First Criterion (Visual Comparison and Dynamic Range)*

To compare these SWI on real images, we pick four regions from Maryland, Oregon, Indiana, and Florida given in Fig. 4.1. The Maryland image (again) represents a typical urban area and contains shadows cast by tall buildings. The Oregon image is of a forested region with tree shadows. There is a small lake in the middle of the Indiana image. Finally, the Florida image is of the ocean. We give the color-coded responses (blue corresponds to the lowest value, and red corresponds to the highest) of the four SWI with respect to each image in Fig. 4.2.

Our first observation is that $\gamma_3$, which uses the green, red, and near-infrared bands, performs poorly in responding to shadows compared to $\gamma_1$, $\gamma_2$, and $\gamma_4$. This is expected since it does not have the blue band in its formation, and we know that the blue band bears the shadow and water information. Nevertheless, $\gamma_3$ represents the best one can expect should only green, red, and near-infrared bands be available.

If we compare $\gamma_2$ with $\gamma_4$, we conclude that $\gamma_2$ outperforms $\gamma_4$ in representing the shadow and water density, while using only the blue, red, and near-infrared bands. So we need just three bands instead of four (as for the $\gamma_4$) to compute it.

Finally, if we compare $\gamma_1$ with $\gamma_2$, for the Oregon image $\gamma_1$ cannot detect shadows on the trees. However, it performs better on the Indiana and Florida images. The main reason for $\gamma_1$'s better performance on the Indiana image is that it responds low to vegetated regions in this image. Finally, for the Maryland image, both $\gamma_1$ and $\gamma_2$ perform well. Based on these observations $\gamma_2$ seems more robust in detecting shadows and water sources.

To compare the dynamic range of these four SWI, we next obtain the normalized histograms on the four images. In Fig. 4.3, we give these histograms in which the

**Fig. 4.1** Sample images to compare the four SWI

(a) The Maryland image

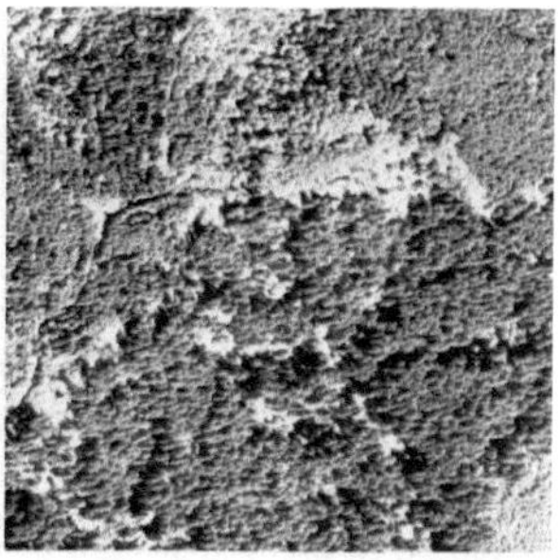

(b) The Oregon image

(c) The Indiana image

(d) The Florida image

**Table 4.1** Comparison of the four SWI on the New Mexico and Florida images

| | $\mu_{\gamma_1}$ | $\mu_{\gamma_2}$ | $\mu_{\gamma_3}$ | $\mu_{\gamma_4}$ |
|---|---|---|---|---|
| The New Mexico image | −0.1896 | 0.1051 | 0.1649 | 0.0904 |
| The Florida image | 0.6939 | 0.5147 | 0.5021 | 0.3895 |

solid curve represents $\gamma_1$, the dotted curve represents $\gamma_2$, the dashed and dotted curve represents $\gamma_3$, and the dashed curve represents $\gamma_4$.

If we compare these histograms, we see that $\gamma_1$ has the highest dynamic range especially on the Oregon and Indiana images. This is again because of its low response to vegetated regions. All other indices have similar dynamic ranges.

Next, let us focus on the two extremes (New Mexico and Florida) to compare responses of the four SWI by their sample means, $\mu$. As we recall, the New Mexico image has neither shadow nor water regions, so the best index response to this image should be a value around zero. The Florida image is of the ocean, so the best index should have the highest response to it. We give these comparisons in Table 4.1.

As seen from the first row of Table 4.1, all indices except $\gamma_3$ performed fairly well when responding to the New Mexico image. The $\gamma_1$ index has a negative response to this image. This is again because of its low response to vegetation. For the Florida image (given in the second row of Table 4.1), $\gamma_1$ outperformed all other indices by having the highest value. $\gamma_2$ and $\gamma_3$ have the second and third highest values. $\gamma_4$ performed poorly. Based on this criterion, we can sort these indices as $\gamma_1$, $\gamma_2$,

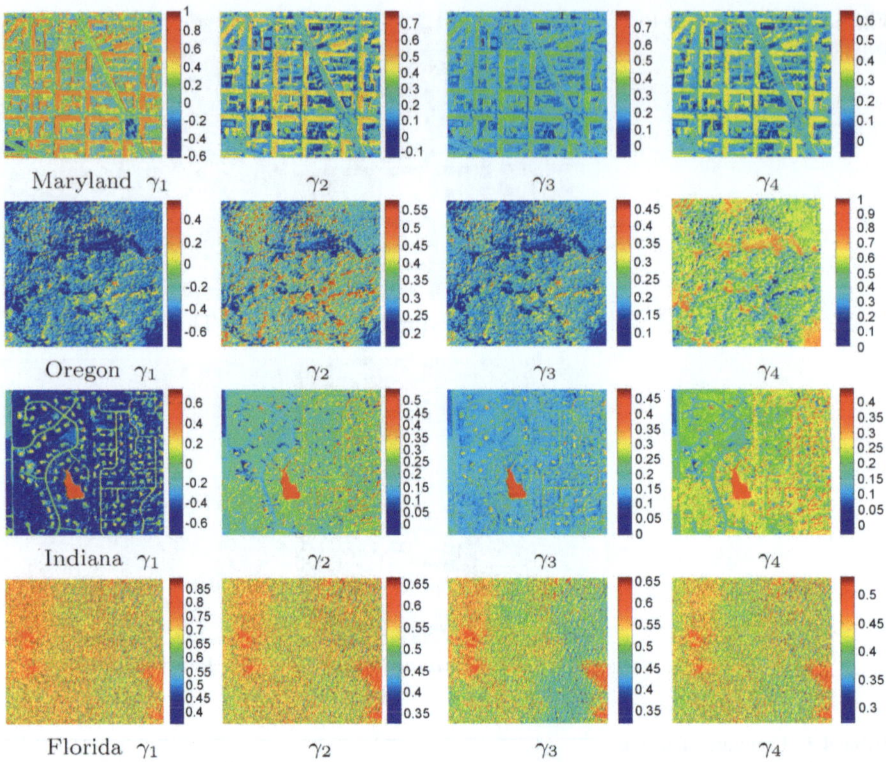


**Fig. 4.2** Color-coded responses of the four SWI to the sample images

$\gamma_3$, and $\gamma_4$ from best to worst. However, $\gamma_1$ has a tendency to respond weakly to vegetated regions, even if they contain shadows.

### *4.1.2 Comparison by the Second Criterion (Entropy on High Contrast Images)*

Next, let us compare these four SWI based on their entropy on the three images containing shadow or water regions. We label images as $img_i$ for $i = 1, 2, 3$ in the order of Maryland, Oregon, and Indiana. We give the results in Table 4.2.

This table indicates that $\gamma_1$ has the highest average entropy on the three images considered. Again this is because of its low response to vegetated regions in these images. $\gamma_2$ has the second highest entropy. $\gamma_4$ and $\gamma_3$ have the third and fourth highest entropies on these images. Based on this criterion, we can sort these indices as $\gamma_2$, $\gamma_4$, and $\gamma_3$ from best to worst, excluding $\gamma_1$ (since its high entropy is, in fact, because of low response to vegetation).

Finally, we conclude that $\gamma_2$ is the most suitable index to measure shadow and water densities in multispectral images. Although $\gamma_1$ outperformed $\gamma_2$ on some ex-

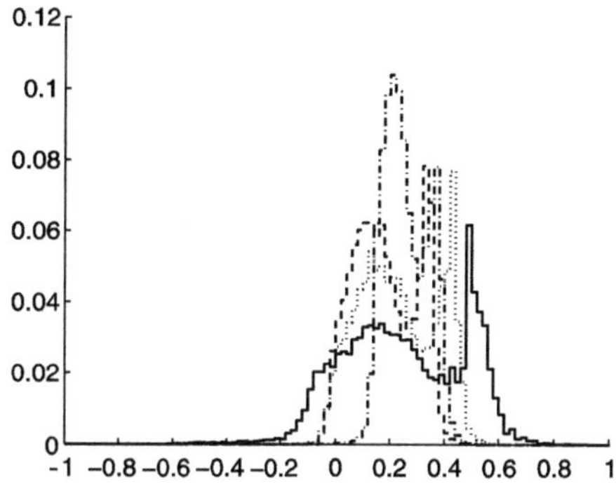


(a) Histograms for the Maryland image

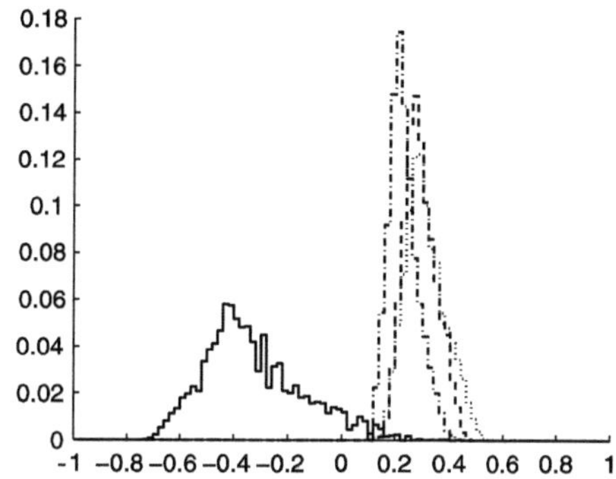


(b) Histograms for the Oregon image

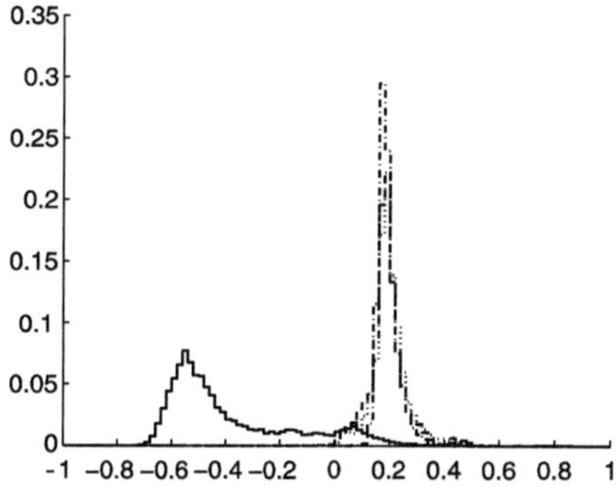


(c) Histograms for the Indiana image

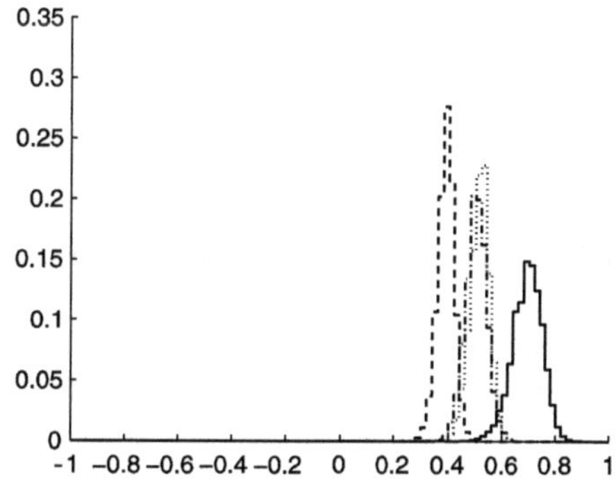


(d) Histograms for the Florida image

**Fig. 4.3** Comparison of histograms for the four SWI to compare their dynamic range. (solid – $\gamma_1$, dot – $\gamma_2$, dash and dot – $\gamma_3$, dash – $\gamma_4$)

**Table 4.2** Comparison of the entropies on three images for the four SWI

| | $img_1$ | $img_2$ | $img_3$ | Mean value |
|---|---|---|---|---|
| $E_{\gamma_1}$ | 5.3427 | 5.1407 | 5.1733 | 5.2189 |
| $E_{\gamma_2}$ | 4.6923 | 3.8475 | 3.5557 | 4.0318 |
| $E_{\gamma_3}$ | 3.9121 | 3.4243 | 2.8545 | 3.3970 |
| $E_{\gamma_4}$ | 4.3985 | 3.5483 | 3.4500 | 3.7989 |

periments, it has a problem in detecting shadows in vegetated regions. Moreover, its superior numerical performance is artificially high due to that low response in vegetated regions. We next give $\gamma_2$ in a simplified form:

$$\gamma_2 = \frac{4}{\pi}\arctan\left(\frac{0.6864\,\rho_{\text{blue}} + 0.7253\,\rho_{\text{red}} + 0.0537\,\rho_{\text{nir}}}{\sqrt{\rho_{\text{blue}}^2 + \rho_{\text{red}}^2 + \rho_{\text{nir}}^2}}\right). \tag{4.5}$$

## 4.2 Summary of the Chapter

In this chapter, we introduced shadow and water indices based on the statistical framework used in the previous chapter. We benefit from these indices to detect water bodies like lakes in residential regions in the following chapters. Although used extensively in computer vision, there has been little work on detecting water regions and shadows from multispectral imagery. Therefore, the indices introduced in this chapter may be of use in computer vision community.

## 4.3 Problems

**4.1** Why do researchers need an index to detect the water regions in a satellite image?

**4.2** Why is separating the shadow and water information hard in multispectral images?

**4.3** Which shadow–water index gives the best result?

**4.4** (**Open ended question**) As in the previous chapter, does using ICA (or another nonlinear transformation) lead to a stronger shadow–water index?

## References

1. V. Klemas, Remote sensing of estuaries: An overview. IEEE Ocean, pp. 302–309 (1986)
2. A. Niedermeier, E. Romaneessen, S. Lehner, IEEE Trans. Geosci. Remote Sens. **38**(5), 2270 (2000)
3. T.J. McIntire, J.J. Simpson, IEEE Trans. Geosci. Remote Sens. **40**(9), 1956 (2002)
4. C. Wang, L. Huang, A. Rosenfeld, Pattern Recogn. Lett. **12**, 55 (1991)
5. A.T.S. Ho, in *Proceedings of IPA* (1997), pp. 848–851
6. M.F. Jasinski, IEEE Trans. Geosci. Remote Sens. **34**(5), 804 (1996)
7. J.J. Simpson, J.R. Stitt, IEEE Trans. Geosci. Remote Sens. **36**(3), 880 (1998)
8. G. Lisens, P. Kempencers, F. Fierens, J. Van Rensbergen, in *Proceedings of IGARSS*, vol. 2 (2000), pp. 834–836
9. J.J. Simpson, Z. Jin, J.R. Stitt, IEEE Trans. Geosci. Remote Sens. **38**(2), 972 (2000)
10. H. Spitzer, R. Franck, M. Kollewe, N. Rega, A. Rothkirch, R. Wiemker, in *Proceedings of IGARSS*, vol. 5 (2001), pp. 2256–2258
11. T.E. Avery, G.L. Berlin, *Fundamentals of Remote Sensing and Airphoto Interpretation*, 5th edn. (Macmillan Co, New York, 1992)

# Part III
# Land Use Classification

# Chapter 5
# Review on Land Use Classification

Land use classification covers a wide range of applications from general land cover determination to specific crop type detection. Many approaches have been developed to infer land usage from satellite and aerial images. Our multispectral satellite image understanding system starts with land use classification. Therefore, it is reasonable to analyze the existing literature on this problem first. To do so, we investigate trends in land use classification between years 1967 and 2002 by reviewing the related literature. We consider the seminal work of Fu et al. [1] to be the beginning of automated land use classification. We did not attempt to cover the whole literature; however, we tried to explore a significant and influential portion of it. Specifically, we focused on feature extraction methods using passive sensors and excluded work on classifiers, neural networks, and fuzzy logic.

To investigate the trends in solving this problem, we reviewed over 90 influential papers published in refereed journals. To identify trends, we grouped papers based on their major contribution. We grouped these papers according to their major contribution. One paper could belong to many groups, but we chose to assign each paper to only one "best" group. We summarize key papers in each group and tabulate each study by the type of image used, geographical location considered, and the average performance obtained. To clarify image types used in the literature, we include a section summarizing remote sensing satellites and airborne equipment.

Unfortunately, the results given in most of the papers we reviewed are based on just one (or a very few) images. It is rarely the case that one finds two papers using the same image(s) to evaluate their methods. For this reason, we were unable to compare the performances of these methods directly. Nevertheless, we report their performances to give the reader some idea of their relative performance. In reporting performance, we used the average classification rate as a benchmark. If the method was applied to several images, we report the mean of the average classification rate for each image.

C. Ünsalan, K.L. Boyer, *Multispectral Satellite Image Understanding*,
Advances in Computer Vision and Pattern Recognition,
DOI 10.1007/978-0-85729-667-2_5, 

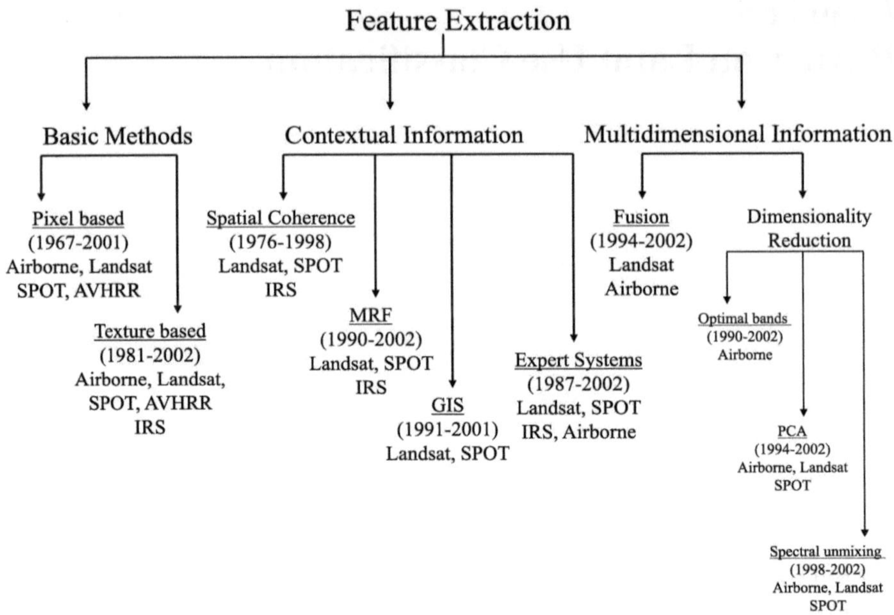


**Fig. 5.1** Taxonomy of the feature extraction methods

## 5.1 Overview of Feature Extraction Methods

Before explaining each feature extraction method in depth, we summarize them in a tree as in Fig. 5.1. Here, we group feature extraction methods into three main categories: basic methods, methods using contextual information, and methods summarizing the multidimensional information. In this figure, we provide the time interval over which each method has been used and the sensors their results are reported on.

In the following sections, we discuss these methods in depth. We also suggest future research directions based on these methods. Let us start with the basic feature extraction methods since they offered an initial approach to solve the land use problem.

## 5.2 Basic Feature Extraction Methods

We consider two basic methods of land use classification in this section. These are based on pixel and texture analysis for feature extraction. These methods represent the earliest approaches to the land use classification problem. In fact, pixel based classification was the first approach explored by researchers. Texture analysis, introduced later, is still used as a powerful feature extraction method. The main similarity between these methods is that most systems based on them are fast and simple compared to the approaches that followed. Unlike later developments, these two approaches do not use spatial information explicitly.

### 5.2.1 Pixel Based Methods

The term "pixel" has a different context in the remote sensing domain than in computer vision and image processing. It can represent a large area with diverse characteristics. For example, AVHRR images capture a 1.21 km$^2$ area in a single pixel. This area may contain many different objects such as buildings, roads segments, and different soil types and covers. Although recent sensors have around 1 m resolution, each pixel may still cover multiple objects or surfaces. For individual object recognition, this may cause problems because the sampling is effectively below the Nyquist limit; however, in land use classification the aim is to classify a region. Thus instead of separate objects, their cumulative (spectral) characteristic is deemed most important. Because a large region is summarized in each pixel, pixel-based methods assume that the gray level or multispectral response at a given pixel suffices for general (average) land use classification.

Fu et al. [1] considered agricultural data in remotely sensed images. Although studies on aerial and satellite images may have appeared prior to this paper, we take it as the first key paper trying to automate the process. Fu et al. used spectral values of pixel values directly as features. They had 12 bands (each taken as a separate feature) in the 0.4–1.0 μm spectral range. To obtain the most representative features, they applied feature selection methods. Finally, they used standard parametric and non-parametric statistical classifiers to obtain the class labels. They also mentioned the importance of spatial information; however, they did not use it in their system.

As a first attempt to solve the land use classification problem, they pointed out all the important steps to be taken, such as feature extraction, feature selection (data dimensionality reduction), classification, and the importance of spatial information. So, this work set the research direction to be followed for the future researchers.

Following Fu's seminal work, there were numerous other pixel based classification studies, for example, [2–8] (Table 5.1). Comparing the performances of these studies directly is not scientifically sound because each uses a different image type and geographical location. But generally, pixel classification rates lie in the mid 80% to mid 90% range. The popularity of pixel based classification methods peaked in the 1980s when image resolution was insufficient for other approaches. Recent pixel

**Table 5.1** Pixel based classification

| Year | Author | Image type | Geo. location | Performance |
|---|---|---|---|---|
| 1967 | Fu et al. [1] | Airborne | Indiana | 94.3% |
| 1980 | Landgrebe [2] | Landsat | Indiana | 94.1% |
| 1987 | Khorram et al. [3] | Landsat | North Carolina | 83.9% |
| 1990 | Etchegorry [4] | Landsat, SPOT | Indonesia | – |
| 1990 | Lloyd [5] | AVHRR | Earth | – |
| 1994 | De Fries and Townshend [6] | AVHRR | New York | 86.8% |
| 1996 | Hoffbeck and Landgrebe [7] | AVIRIS | Nevada | 96.4% |
| 2001 | Bandyopadhyay et al. [8] | SPOT | India | 88.9% |

based classification algorithms use additional information, setting them apart from the initial studies.

### 5.2.2 Texture Analysis Based Methods

The next group of papers frame the problem in terms of texture analysis. Since satellite and aerial images contain many objects while the main objective is to classify regions, researchers have turned to texture analysis methods. Texture analysis has been studied in the computer vision literature for many years. Several "definitions" of texture have been formulated by researchers, and there is no general consensus that can be formalized mathematically. Some definitions are perceptually motivated, others are driven completely by the application at hand. One early and influential definition is due to Haralick:

"The image texture we consider is non-figurative and cellular... An image texture is described by the number and types of its (tonal) primitives and the spatial organization or layout of its (tonal) primitives... A fundamental characteristic of texture: It cannot be analyzed without a frame of reference of tonal primitive being stated or implied. For any smooth gray-tone surface, there exists a scale such that when the surface is examined, it has no texture. Then as resolution increases, it takes on a fine texture and then a coarse texture." [9].

For the interested reader, we recommend two excellent review papers by Tuceryan and Jain [10] and Wezska et al. [11]. In nearly all texture analysis methods used for land use classification, an operation performed on an image region produces a representative number for use in classification.

Irons and Petersen [12] applied texture analysis methods to a Landsat image of Clearfield County, Pennsylvania, using the mean, variance, skewness, kurtosis, and the mean of the maximum gray level difference as features. They conclude that these first-order statistical features are not very useful for land use classification. Although their results were not encouraging, applying texture analysis as a feature extraction method in land use classification was the main contribution of their paper.

In another study, Conners et al. [13] applied the Grey Level Cooccurrence Matrix (GLCM) features to a Sunnyvale, California, image. GLCM features are extracted based on (estimated) conditional probabilities of neighboring grey level values [14, 15]. These features are second-order statistics. Being a powerful feature extraction method in texture analysis, GLCM has been applied extensively in land use classification. Conners et al. were the first to provide this method to solve the land use classification problem.

Many researchers have applied texture analysis methods to land use classification [16–24] (Table 5.2). As Table 5.2 shows, the classification rates are in the 80–90% range. These methods grew in popularity during the late 1990s as the resolution of satellite images increased and simple pixel based classification became obsolete.

**Table 5.2** Texture analysis based classification

| Year | Author | Image type | Geo. location | Performance |
|---|---|---|---|---|
| 1981 | Irons and Petersen [12] | Landsat | Pennsylvania | – |
| 1984 | Conners et al. [13] | Airborne | California | 83.4% |
| 1990 | Marceau et al. [16] | SPOT | Canada | 80.0% |
| 1991 | Lee and Philpot [17] | Landsat | New York | 80.0% |
| 1992 | Sali and Wolfson [18] | SPOT | New York | – |
| 1995 | Baraldi and Parmiggiani [19] | AVHRR | Antarctica | – |
| 1999 | Smits and Annoni [20] | IRS | Italy | 90.0% |
| 2000 | Karathanassi et al. [21] | SPOT | Greece | 89.6% |
| 2001 | Shaban and Dikshit [22] | SPOT | India, Egypt | 80.2% |
| 2001 | Cihlar et al. [23] | Landsat | Canada | – |
| 2002 | Parinello and Vaughan [24] | AVHRR | Scotland | – |

## 5.3 Methods Using Contextual Information

Another major category of research focuses on the use of contextual information in land use classification. We divided these studies into four subgroups: spatial coherence, Markov Random Fields (MRF), geographic information systems, and expert systems. These methods account for spatial interactions between neighboring pixels and regions in classification to improve performance.

### *5.3.1 Spatial Coherence*

When seen from overhead, similar land types tend to be more or less contiguous, that is, neighboring pixels have a high probability of belonging to the same land class; pixels lying on a border between two different classes are relatively rare. This observation has led researchers to exploit spatial coherence in land use classification.

Kettig and Landgrebe [25] were among the first researchers to use this constraint. They grouped contiguous pixels into homogeneous sets and then classified the groups. They applied their algorithm to two Landsat and two airborne images and observed an improvement in the classification performance. Their paper provides as a key contribution in the use of spatial coherence.

Baraldi and Parmiggiani [26] used a SPOT image from Modena, Italy, to test a classifier for urban areas, using textural features and a region growing algorithm as the contextual classifier. They concluded that the contextual classifier was performing as desired, but that textural features are not satisfactory for SPOT imagery. Using region growing as a grouping method for land use classification was the unique feature of this paper.

**Table 5.3** Classification using spatial information

| Year | Author | Image type | Geo. location | Performance |
|---|---|---|---|---|
| 1976 | Kettig and Landgrebe [25] | Landsat | – | – |
| 1980 | Thomas [30] | Landsat | New Zealand | – |
| 1980 | Fu and Yu [31] | Landsat | Indiana | 89.2% |
| 1982 | Tilton et al. [32] | Landsat | Kansas, Indiana | 86.0% |
| 1982 | Richards et al. [28] | – | – | 81.0% |
| 1982 | Wharton [33] | Simulation | – | 88.0% |
| 1983 | Gurney and Townshend [34] | Landsat | UK | 70.0% |
| 1987 | Cushnie [35] | – | – | – |
| 1990 | Khazenie and Crawford [36] | Simulation | – | 87.3% |
| 1990 | Jensen [27] | Landsat | Bangkok | – |
| 1990 | Kusaka et al. [37] | SPOT | Japan | 78.0% |
| 1990 | Baraldi and Parmiggiani [26] | SPOT | Italy | – |
| 1991 | Kim and Crawford [38] | SPOT | Texas | – |
| 1991 | Alonso and Soria [39] | Landsat | Spain | 70.4% |
| 1992 | Gong and Howarth [40] | SPOT | Canada | 61.6% |
| 1992 | Gong and Howarth [41] | SPOT | Canada | 72.2% |
| 1996 | Barnsley and Barr [42] | SPOT | England | 97.1% |
| 1997 | Lobo [43] | Landsat | Spain | 84.5% |
| 1998 | Kertikeyan et al. [44] | IRS | – | 91.0% |
| 1998 | Sharma and Sarkar [45] | Landsat | India | 89.5% |

Jensen [27] used the contextual classification idea with the GLCM features. In this way he combined both methods. He tested on a 30 meter resolution Landsat image of Bangkok and gave a visual interpretation. Unfortunately, this interpretation is inconclusive for assessing the power of the method. However, combining the GLCM features with contextual classification was the extension, this paper offered.

Richards et al. [28] applied probabilistic relaxation to inject spatial information into the classification process. Probabilistic relaxation is a labeling method taking into account interactions between neighboring pixels [29]. Richards et al. first obtained the Maximum Likelihood (ML) classification labeling and applied relaxation to these labels to include spatial coherence. Using relaxation (which proved its power in a wide range of applications) instead of other spatial information processing systems was the key contribution of this paper.

Other work in this area includes [30–45] (Table 5.3). As Table 5.3 shows, the classification rate for these methods lie in the range of 70% to 97%. Although most popular in the mid-1990s, they survive in different forms, as we mention next.

**Table 5.4** MRF based classification

| Year | Author | Image type | Geo. location | Performance |
|---|---|---|---|---|
| 1990 | Zhang et al. [47] | Landsat | Virginia | 77.0% |
| 1992 | Jeon and Landgrebe [48] | Landsat | Indiana | 82.8% |
| 1996 | Jhung and Swain [49] | Landsat | Indiana | 86.2% |
| 1996 | Solberg et al. [50] | Landsat | Norway | 96.9% |
| 1999 | Yamazaki and Gingras [51] | Landsat | Canada | 78.2% |
| 1999 | Tso and Mather [52] | Landsat | Sudan | 81.0% |
| 1999 | Kerfoot and Bresler [53] | Landsat | Illinois | – |
| 1999 | Yu et al. [54] | SPOT | France | – |
| 2002 | Sarkar et al. [55] | IRS | – | 95.0% |

### *5.3.2 Markov Random Fields*

Another popular tool to imbed contextual information into classification is the Markov Random Fields (MRF). This method models the joint statistics of a small neighborhood of pixels. Its popularity for texture analysis initiated its application to land use classification. The seminal work of Geman and Geman [46] provides an excellent development of this method. We consider MRFs to be part of the continuum of spatial information based methods.

The method can be briefly summarized as follows. The brightness level at an image point is highly dependent on (and therefore correlated with) the brightness level of neighboring points, assuming the image is of a real coherent scene and not merely random noise. The MRF provides a precise model for this dependence.

Zhang et al. [47] introduced the MRF as a contextual classification algorithm. They captured the correlation between neighboring pixels by MRF models then applied stochastic relaxation to obtain the final class labels. They applied this method on subwindows of test images with four spectral bands. Zhang's work showed the usefulness of MRF modeling in capturing the context information for land use classification problem. Different from the spatial information, this method used stochastic modeling via MRF.

Other researchers also used MRFs in their systems [48–55] (Table 5.4). As Table 5.4 shows, the classification performance for the MRF-based systems ranges from 77% to 97%. Publication dates indicate that MRF based methods gained acceptance in the late 1990s.

### *5.3.3 Geographical Information Systems*

Another class of methods to exploit contextual information are those using Geographical Information Systems (GIS) and topological maps in classification. In this case, the contextual information derives from prior knowledge of a specific region,

**Table 5.5** GIS

| Year | Author | Image type | Geo. location | Performance |
|---|---|---|---|---|
| 1991 | Pedley and Curran [56] | SPOT | – | 62.1% |
| 1993 | Wang [57] | Landsat | Canada | 83.8% |
| 2000 | De Bruin and Gorte [58] | Landsat | Spain | 79.5% |
| 2000 | De Bruin [59] | Landsat | Spain | – |
| 2001 | Baban and Yusof [60] | Landsat | Malaysia | 92.0% |

rather than generic observations regarding spatial coherence in land use. As such, these systems are geared to land use updating rather than raw classification.

One of the first applications of this idea is by Pedley and Curran [56]. They grouped pixels according to the GIS data and classified these groups. Compared to pixel classification, Pedley and Curran obtained improved performance. Since recently GIS has been used extensively, Pedley and Curran's work can be taken one of the most effective tools in updating land use classification. However, their method requires a prior classification which limits its application.

The work in this area includes [57–60] (Table 5.5). As Table 5.5 shows, the classification performance ranges from 62% to 92% for these methods. Although researchers started using these methods around 1991 they are becoming popular nowadays as GIS databases become more prevalent and populated. Indeed, we expect most future systems—using all available prior contextual information and spatial coherence—to incorporate such databases, as well.

### *5.3.4 Expert Systems*

Expert systems cover a large range of methods such as artificial intelligence, rule based systems, knowledge based representations, Bayesian belief networks, and case based reasoning. Jackson's text [61] represents a good introductory reference on this subject.

There have been numerous studies in land use classification using expert systems in which the rule base encapsulates the contextual information. To our knowledge, the initial work of Wharton [62] in 1987 was the first study to use expert systems in land use classification (but not in remote sensing).

He considered spectral values of pixels (as in previous studies) as features. He then used a knowledge based approach to capture the spectral relationships between classes. To assign the final class labels, he iteratively updated class labels based on the knowledge base till the class labels stabilize (till most of the pixels satisfy the neighborhood rules).

Wharton's approach was another perspective for contextual information processing. Unlike previous three approaches, he captured the context information by defining rules on the neighborhood regions. This way, his approach takes the perspective of artificial intelligence.

**Table 5.6** Expert systems

| Year | Author | Image type | Geo. location | Performance |
|---|---|---|---|---|
| 1980 | Nagao and Matsuyama [70] | Airborne | Japan | – |
| 1987 | Wharton [62] | Landsat | Maryland | 85.0% |
| 1991 | Ton et al. [63] | Landsat | Michigan | 86.0% |
| 1995 | Kartikeyan et al. [64] | IRS | – | 92.0% |
| 1997 | Jensen [65] | Landsat, SPOT | Ghana | – |
| 2000 | Casasnovas [66] | Landsat | Spain | 81.4% |
| 2000 | Barr and Barnsley [67] | Daedalus | UK | 82.4% |
| 2001 | Stefanov et al. [68] | Landsat | Arizona | 85.0% |
| 2002 | Datcu et al. [69] | Daedalus | UK, Germany | 88.8% |

Ton et al. [63] also used a complex knowledge-based system to classify Landsat images. They extracted major land types using a hierarchical classifier, followed by a rule-based clustering algorithm to obtain fine classification. They obtained promising results, albeit with a limited test set. A rule based clustering algorithm is also another powerful approach to obtain the final class labels.

Other research in this area includes [64–69] (Table 5.6). Table 5.6 indicates the increasing popularity of these systems with their performance ranging from 81% to 92%.

## 5.4 Methods Summarizing Multidimensional Information

The following papers focus on the dimensionality of the data to help (or replace) feature extraction and the inclusion of spatial information. Either by reducing the dimensionality of the data or by fusing different representations of the same pixel, they aim to improve performance. We divide these papers into two subgroups: data dimensionality reduction and data-decision fusion.

### *5.4.1 Data Dimensionality Reduction*

In land use classification, the same scene is often imaged at multiple wavelengths (bands). For multispectral images this representation is around four to seven bands; hyperspectral images may go up to 240 bands (dimensions). Therefore, each pixel in these images is represented as a multidimensional vector, having four to 240 entries. Not all of these bands are equally useful for classification, depending on the application. Therefore, the data dimension may be reduced either by selecting useful bands (optimal band selection) or by transforming the original space to one of lower dimension (principal components analysis and spectral unmixing).

**Table 5.7** Optimal band selection

| Year | Author | Image type | Geo. location | Performance |
|---|---|---|---|---|
| 1990 | Mausel et al. [71] | Airborne | Texas | 92.2% |
| 2002 | Tsai and Philpot [72] | AVIRIS | California | 69.0% |

**Table 5.8** PCA

| Year | Author | Image type | Geo. location | Performance |
|---|---|---|---|---|
| 1994 | Harsyani and Chang [74] | AVIRIS | Nevada | – |
| 1999 | Jia and Richards [75] | AVIRIS | California | 97.8% |
| 2000 | Kuplich et al. [76] | Landsat | Brazil | 94.8% |
| 2000 | Ren and Chang [77] | Landsat, SPOT | Virginia | – |
| 2001 | Kumar et al. [78] | AVIRIS | Florida | 95.0% |

#### 5.4.1.1 Optimal Band Selection

Mausel et al. [71] compared four optimum band selection methods for land use classification. Their paper serves as a good reference for researchers planning to use these band selection techniques. Unlike dimension reduction by transforming the original space, these methods serve as feature selection criteria in the original space. Tsai and Philpot's [72] paper is another work on optimal band selection (Table 5.7).

#### 5.4.1.2 Principal Components Analysis

Principal Components Analysis (PCA) performs a linear transformation on a set of (possibly) correlated random vectors to represent them in a new space such that they are uncorrelated. The new coordinate space is related to the original by a rotation about the origin, such that the new coordinate axes are aligned with the dominant directions of dispersion (the principal components) of the data, as inferred from its covariance matrix. For a thorough treatment, see [73].

Harsyani and Chang [74] considered an extended version of the PCA to reduce data dimensionality. As we recall, in hyperspectral images a pixel may be represented by (up to) a 240 dimensional vector. For such high dimensional data, PCA and similar transformations are needed for reliable classification. Harsyani and Chang aimed to reduce the data dimensionality while retaining most of the discriminating information via their method. This paper pointed to another powerful research area for data dimension reduction. Similar studies on PCA and related transformation methods include [75–78] (Table 5.8).

**Table 5.9** Spectral unmixing

| Year | Author | Image type | Geo. location | Performance |
|---|---|---|---|---|
| 1998 | Chang et al. [79] | AVIRIS | Nevada | – |
| 1999 | Chang and Brumbley [80] | SPOT | USA | – |
| 2000 | Chang and Ren [81] | HYDICE | Maryland | 95.0% |
| 2001 | Small [82] | Landsat | New York | – |
| 2001 | Roessner et al. [83] | DAIS 7915 | Germany | 80.0% |
| 2002 | Chang [84] | AVIRIS, HYDICE | Nevada | – |
| 2002 | Verhoeye and De Wulf [85] | SPOT | Nigeria | 91.0% |

#### 5.4.1.3 Spectral Unmixing

To overcome problems in PCA, many researchers considered spectral unmixing techniques. Chang et al. [79] describe spectral unmixing as:

"Linear spectral unmixing is widely used approach in multispectral/hyperspectral imagery to determine and quantify individual spectral signatures in a mixed pixel."

Chang et al. also showed the usefulness of spectral unmixing in land use classification. Papers using spectral unmixing methods include [80–85] (Table 5.9).

As we consider Tables 5.7, 5.8, and 5.9, we notice that data dimension reduction has a positive effect on classification performance. The publication dates suggest that these methods are growing more popular.

### *5.4.2 Data and Decision Fusion*

In decision fusion, decisions of different classifiers are combined in such a way that the final classification performance is improved. There have been many methods proposed for this purpose, and an excellent review paper on this subject is by Kittler et al. [86].

Instead of data dimension reduction, all of the bands are used in decision fusion. However, classification is applied to each band separately, and the assigned class labels are fused. In a way, this is the dual of data dimension reduction, in that the order of dimension reduction and classification is replaced.

Jeon and Landgrebe [87] proposed two decision fusion based classifiers. With their method they reported a 5% improvement in performance. This improvement itself indicates the usefulness of the idea.

In a related study, Solberg et al. [88] applied data fusion on pixel basis and obtained superior performance. The goal of data fusion is to obtain a useful representation from the multidimensional data set. Clark and Yuille's book [89] is a very good reference on this subject.

Decision fusion have been studied by many authors in land use classification [90–94] (Table 5.10). As Table 5.10 shows, these methods have been considered since 1994 and most studies are quite new (by looking at their publication dates).

**Table 5.10** Decision fusion

| Year | Author | Image type | Geo. location | Performance |
|---|---|---|---|---|
| 1994 | Solberg et al. [88] | Landsat | Norway | 95.0% |
| 1997 | Benediktsson et al. [90] | Landsat | Colorado | 75.0% |
| 1999 | Jeon and Landgrebe [87] | Landsat | Indiana | 83.6% |
| 1999 | Bennediktsson et al. [91] | Landsat | Portugal, Iceland | 86.0% |
| 1999 | Jimenez et al. [92] | AVIRIS | Indiana | 96.0% |
| 2000 | Steele [93] | Landsat | Idaho, Montana | 68.0% |
| 2002 | Smits [94] | Landsat | Netherlands | 95.0% |

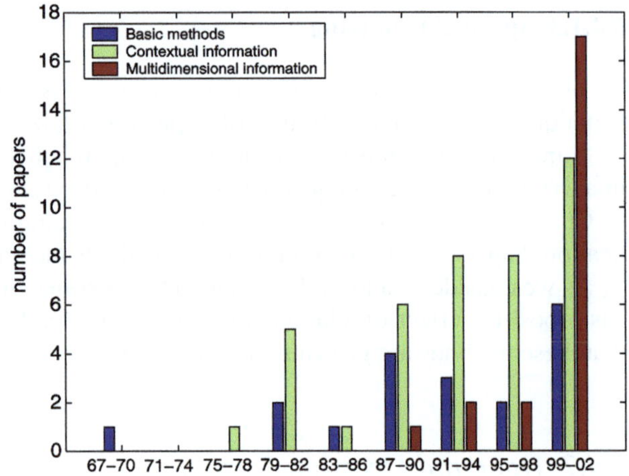


**Fig. 5.2** Number of papers published in three year time intervals from 1967 to 2002

## *5.4.3 Summary of the Methods*

In the previous sections, we grouped studies in land use classification to three main categories. Initial research in the area started with basic methods [1, 13]. The next class of studies dealt with contextual information [47, 62]. The last group of studies dealt with data dimension reduction and decision fusion [74, 79, 87]. These key papers summarize the main trends in land use classification since 1967. Next, we plot the number of papers published in three year intervals from 1967 to 2002 in Fig. 5.2 to summarize the popularity of these methods.

Figure 5.2 points to the following conclusions. After their introduction, methods on contextual information dominated basic methods. This indicates the importance of contextual information in land use classification. Although introduced later, work on multidimensional information has recently taken the lead in the number of papers published. This indicates that the main trend nowadays is towards multidimensional information processing. The next most promising direction is in contextual information processing. If we consider these main groups in detail, we observe the following.

Within basic methods, texture analysis is becoming more popular than pixel based techniques. If we check the usage of sensors, almost all of them have been used for basic methods category. One important point is that AVHRR finds use in basic methods only. Its limited resolution renders it unsuitable for the more sophisticated techniques that have followed.

Methods using contextual information can be sorted based on their recent publication dates into: expert systems, GIS, MRF, and spatial coherence. In this category expert systems are becoming more popular. As we mentioned, GIS based systems are likely to take the lead from expert systems in the near future. Sensors used in this category are dominantly Landsat and SPOT. The properties of these sensors seem suitable for capturing the context.

We can sort the methods summarizing the multidimensional information by their recent publication dates into: spectral unmixing, PCA, and data and decision fusion. Spectral unmixing, proposed to overcome the shortcomings of PCA is popular nowadays, and PCA based data dimension reduction techniques follow. Sensors used in this category are dominantly airborne for data dimension reduction and Landsat for decision fusion. Since data dimension reduction techniques are proposed for high dimensional data, it is self-evident that they find use with airborne sensors producing hyperspectral data. As for decision fusion, Landsat seems to be the most suitable sensor based on the current publications in this area.

## 5.5 Summary of the Chapter

In this chapter, we identified dominant trends in land use classification by reviewing a large and influential portion of the literature. We also considered publication dates in these major research areas to gain an idea of the life span of each method. Here, we try to project our observations to estimate the future research directions.

As we mentioned in Chap. 2, the resolution of satellite and airborne sensors is increasing. Commercially available satellite data have around 0.45 m resolution per pixel. Therefore, the basic assumption in pixel based land use classification no longer holds. So, basic pixel based feature extraction is not the most effective use of these sensors. Since we still observe texture on these commercially available satellite data, texture analysis may still be feasible for land use classification. Texture analysis as a feature extraction method in decision fusion is a promising research area to be explored.

As for systems using contextual information, Bayesian networks [95] and perceptual inference networks [96] can be considered for future research. Both methods take context information into account and proved to be useful for other research areas. We introduced a feature based grouping method (perceptual organization) as a spatial information processing method [97]. This method has a good performance. In the following chapters we will investigate it further.

For multidimensional information processing, there are many possible research areas to be explored. One area is applying projection pursuit to data dimensionality reduction. Ifarraguerri and Chang [98] studied this approach for general remote sensing applications. It may also be used as a powerful tool in data dimensionality reduction. Similarly, independent components analysis [99] can be used for data dimension reduction.

One other fruitful approach may be using decision fusion with contextual information in succession. Similarly, applying data dimension reduction, followed by decision fusion on the remaining bands or band combinations may be fruitful.

Besides the existing categories, one emerging new category is the usage of structural information in land use classification. Up to now, structural approaches have not been considered by researchers. We consider these methods in the following chapters.

Satellite image resolution, especially images from Ikonos, permit the detection of objects (such as houses and road networks) in the scene. This allows researchers to infer surface characteristics directly from the objects and to classify regions based on this information. This, we believe, will lead to the extensive use of techniques from computer vision in land use classification. We will cover a set of techniques on this problem in the following chapters.

Research on land use classification is still growing, and the field continues to mature. With different perspectives it offers great promise to researchers in the future. Hopefully, this review chapter will be of assistance to researchers working on this problem, by serving as a guide for the past achievements and suggesting possible avenues for future studies.

## 5.6 Problems

**5.1** Based on the information provided in this chapter, discuss the relationship between the image resolution and the land use classification method used.

**5.2** Why is land use classification important for researchers?

**5.3** What are the basic trends for land use classification?

**5.4** Using the same taxonomy in this chapter, add recent research papers on land use classification.

**5.5 (Open ended question)** Which land use classification method seems most promising for future applications?

## References

1. K.S. Fu, D.A. Landgrebe, T.L. Philips, Proc. IEEE **57**, 639 (1969)
2. D.A. Landgrebe, Pattern Recognit. **12**, 165 (1980)
3. S. Khorram, J.A. Brockhaus, H.M. Cheshire, IEEE Trans. Geosci. Remote Sens. **25**(2), 238 (1987)

4. J.P. Gastellu-Etchegorry, Int. J. Remote Sens. **11**, 225 (1990)
5. D. Lloyd, Int. J. Remote Sens. **11**, 2269 (1990)
6. R.S. DeFries, J.G.R. Townshend, Int. J. Remote Sens. **15**, 3567 (1994)
7. J.P. Hoffbeck, D.A. Landgrebe, Remote Sens. Environ. **57**, 119 (1996)
8. S. Bandyopadhyay, S.K. Pal, IEEE Trans. Geosci. Remote Sens. **39**(2), 303 (2001)
9. R.M. Haralick, Statistical and structural approaches to texture. Proceedings of the IEEE, pp. 786–804 (1979)
10. M. Tuceryan, A.K. Jain, in *The Handbook of Pattern Recognition and Computer Vision* (1993), pp. 235–276
11. J.S. Weszka, C.R. Dyer, A. Rosenfeld, IEEE Trans. Syst. Man Cybern. **6**(4), 269 (1976)
12. J.R. Irons, G.W. Petersen, Remote Sens. Environ. **11**, 359 (1981)
13. R.W. Conners, M.A. Trivedi, C.A. Harlow, Comput. Vis. Graph. Image Process. **25**, 273 (1984)
14. R.M. Haralick, K. Shanmugam, IEEE Trans. Geosci. Electron. **11**, 171 (1973)
15. R.M. Haralick, K. Shanmugam, I. Dinstein, IEEE Trans. Syst. Man Cybern. **3**, 610 (1973)
16. D.J. Marceau, P.J. Howarth, J.M.M. Dubois, D.J. Gratton, IEEE Trans. Geosci. Remote Sens. **28**(4), 513 (1990)
17. J.H. Lee, W.D. Philpot, IEEE Trans. Geosci. Remote Sens. **29**(4), 545 (1991)
18. E. Sali, H. Wolfson, Int. J. Remote Sens. **13**, 3395 (1992)
19. A. Baraldi, F. Parmiggiani, IEEE Trans. Geosci. Remote Sens. **33**(2), 293 (1995)
20. P.C. Smits, A. Annoni, IEEE Trans. Geosci. Remote Sens. **37**(3), 1244 (1999)
21. V. Karathanassi, C. Iossifidis, D. Rokos, Int. J. Remote Sens. **21**(9), 1807 (2000)
22. M.A. Shaban, O. Dikshit, Int. J. Remote Sens. **22**(4), 565 (2001)
23. J. Cihlar, J. Okouneva, G. Beaubien, R. Latifovic, Int. J. Remote Sens. **22**(11), 2151 (2001)
24. T. Parrinello, R.A. Vaughan, Int. J. Remote Sens. **23**(9), 1799 (2002)
25. R.L. Kettig, D.A. Landgrebe, IEEE Trans. Geosci. Electron. **14**(1), 19 (1976)
26. A. Baraldi, F. Parmiggiani, IEEE Trans. Geosci. Remote Sens. **28**(4), 674 (1990)
27. L. Moller-Jensen, Photogramm. Eng. Remote Sens. **56**, 899 (1990)
28. J.A. Richards, D.A. Landgrebe, P.H. Swain, Remote Sens. Environ. **12**, 463 (1982)
29. A. Rosenfeld, R.A. Hummel, S.W. Zucker, IEEE Trans. Syst. Man Cybern. **6**, 420 (1973)
30. I.L. Thomas, Photogramm. Eng. Remote Sens. **46**, 1201 (1980)
31. K.S. Fu, T.S. Yu, *Statistical Pattern Classification Using Contextual Information* (Research Studies Press, USA, 1980)
32. J.C. Tilton, S.B. Vardeman, P.H. Swain, IEEE Trans. Geosci. Electron. **20**, 445 (1982)
33. S.W. Wharton, Pattern Recognit. **15**, 317 (1982)
34. C.M. Gurney, J.R.G. Townshend, Photogramm. Eng. Remote Sens. **49**, 55 (1983)
35. J.L. Cushnie, Int. J. Remote Sens. **8**(1), 15 (1987)
36. N. Khazenie, M.M. Crawford, IEEE Trans. Geosci. Remote Sens. **28**(4), 529 (1990)
37. T. Kusaka, H. Egava, Y. Kawata, IEEE Trans. Geosci. Remote Sens. **28**(4), 749 (1990)
38. K. Kim, M.M. Crawford, IEEE Trans. Geosci. Remote Sens. **29**(4), 481 (1991)
39. F.G. Alonso, S.L. Soria, Int. J. Remote Sens. **12**, 2227 (1991)
40. P. Gong, P.J. Howarth, Photogramm. Eng. Remote Sens. **58**, 423 (1992)
41. P. Gong, P.J. Howarth, Int. J. Remote Sens. **13**, 1459 (1992)
42. M.J. Barnsley, S.L. Barr, Photogramm. Eng. Remote Sens. **62**, 949 (1996)
43. A. Lobo, IEEE Trans. Geosci. Remote Sens. **35**(5), 1136 (1997)
44. B. Kartikeyan, A. Sarkar, K.L. Majumder, Int. J. Remote Sens. **19**, 1695 (1998)
45. K.M.S. Sharma, A. Sarkar, Photogramm. Eng. Remote Sens. **64**, 273 (1998)
46. S. Geman, D. Geman, IEEE Trans. Pattern Anal. Mach. Intell. **6**(6), 721 (1984)
47. M.C. Zhang, R.M. Haralick, J.B. Campbell, IEEE Trans. Syst. Man Cybern. **20**(1), 128 (1990)
48. B. Jeon, D.A. Landgrebe, IEEE Trans. Geosci. Remote Sens. **30**(4), 663 (1992)
49. Y. Jhung, P.H. Swain, IEEE Trans. Geosci. Remote Sens. **34**(1), 67 (1996)
50. A.H.S. Solberg, T. Taxt, A.K. Jain, IEEE Trans. Geosci. Remote Sens. **34**(1), 100 (1996)
51. T. Yamazaki, D. Gingras, IEEE Trans. Geosci. Remote Sens. **37**(2), 1173 (1999)
52. B.C.K. Tso, P.M. Mather, IEEE Trans. Geosci. Remote Sens. **37**(3), 1255 (1999)

53. I.B. Kerfoot, Y. Bresler, IEEE Trans. Image Process. **8**(6), 798 (1999)
54. S. Yu, M. Berthod, G. Giraudon, IEEE Trans. Geosci. Remote Sens. **37**(4), 1925 (1999)
55. A. Sarkar, K. Biswas, B. Kartikeyan, V. Kumar, K.L. Majumder, D.K. Pal, IEEE Trans. Geosci. Remote Sens. **40**(5), 1102 (2002)
56. M.I. Pedley, P.J. Curran, Int. J. Remote Sens. **12**, 2181 (1991)
57. F. Wang, IEEE Trans. Geosci. Remote Sens. **31**(1), 136 (1993)
58. S.D. Bruin, B.G.H. Gorte, Int. J. Remote Sens. **21**(12), 2389 (2000)
59. S.D. Bruin, Remote Sens. Environ. **74**, 387 (2000)
60. S.M.J. Baban, K.W. Yusof, Int. J. Remote Sens. **22**(10), 1909 (2001)
61. P. Jackson, *Introduction to Expert Systems*, 3rd edn. (Addison-Wesley, Massachusetts, 1999)
62. S.W. Wharton, IEEE Trans. Geosci. Remote Sens. **25**(3), 272 (1987)
63. J. Ton, J. Sticklen, A.K. Jain, IEEE Trans. Geosci. Remote Sens. **29**(2), 222 (1991)
64. B. Kartikeyan, K.L. Majumder, A.R. Dasgupta, IEEE Trans. Geosci. Remote Sens. **33**(1), 58 (1995)
65. L.M. Jensen, Comput. Environ. Urban Syst. **21**(3/4), 291 (1997)
66. J.A.M. Casasnovas, Int. J. Remote Sens. **21**(9), 1825 (2000)
67. S. Barr, M. Barnsley, Comput. Geosci. **26**, 433 (2000)
68. W.L. Stefanov, M.S. Ramsey, P.R. Christensen, Remote Sens. Environ. **77**, 173 (2001)
69. M. Datcu, F. Melgani, A. Piardi, S.B. Serpico, IEEE Trans. Geosci. Remote Sens. **40**(3), 609 (2002)
70. M. Nagao, T. Matsuyama, *A Structural Analysis of Complex Aerial Photographs* (Plenum, New York, USA, 1980)
71. P.W. Mausel, W.J. Kramber, J.K. Lee, Photogramm. Eng. Remote Sens. **56**, 55 (1990)
72. F. Tsai, W.D. Philpot, IEEE Trans. Geosci. Remote Sens. **40**(2), 416 (2002)
73. I.T. Joliffe, *Principal Component Analysis*, 2nd edn. (Springer, New York, 2002)
74. J.C. Harsanyi, C.I. Chang, IEEE Trans. Geosci. Remote Sens. **32**(4), 779 (1994)
75. X. Jia, J.A. Richards, IEEE Trans. Geosci. Remote Sens. **37**(1), 538 (1999)
76. T.M. Kuplich, C.C. Freitas, J.V. Soares, Int. J. Remote Sens. **21**(10), 2101 (2000)
77. H. Ren, C.I. Chang, IEEE Trans. Geosci. Remote Sens. **38**(6), 2515 (2000)
78. S. Kumar, J. Ghosh, M.M. Crawford, IEEE Trans. Geosci. Remote Sens. **39**(7), 1368 (2001)
79. C.I. Chang, X.L. Zhao, M.L.G. Althouse, J.J. Pan, IEEE Trans. Geosci. Remote Sens. **36**(3), 898 (1998)
80. C.I. Chang, C.M. Brumbley, IEEE Trans. Geosci. Remote Sens. **37**(1), 257 (1999)
81. C.I. Chang, H. Ren, IEEE Trans. Geosci. Remote Sens. **38**(2), 1044 (2000)
82. C. Small, Int. J. Remote Sens. **22**(7), 1305 (2001)
83. S. Roessner, U. Heiden, H. Kaufmann, IEEE Trans. Geosci. Remote Sens. **39**(7), 1525 (2001)
84. C.I. Chang, IEEE Trans. Geosci. Remote Sens. **40**(5), 1065 (2002)
85. J. Verhoeye, R.D. Wulf, Remote Sens. Environ. **79**, 96 (2002)
86. J.V. Kittler, M. Hatef, R.P.W. Duin, J. Matas, IEEE Trans. Pattern Anal. Mach. Intell. **20**(3), 226 (1998)
87. B. Jeon, D.A. Landgrebe, IEEE Trans. Geosci. Remote Sens. **37**(3), 1227 (1999)
88. A.H.S. Solberg, A.K. Jain, T. Taxt, IEEE Trans. Geosci. Remote Sens. **32**(4), 768 (1994)
89. J.J. Clark, A. Yuille, *Data Fusion for Sensory Information Processing Systems* (Kluwer Academic, Boston, 1990)
90. J.A. Benediktsson, J.R. Sveinsson, P.H. Swain, IEEE Trans. Geosci. Remote Sens. **35**(4), 833 (1997)
91. J.A. Benediktsson, I. Kanellopoulos, IEEE Trans. Geosci. Remote Sens. **37**(3), 1367 (1999)
92. L.O. Jimenez, A.M. Morell, A. Creus, IEEE Trans. Geosci. Remote Sens. **37**(3), 1360 (1999)
93. B.M. Steele, Remote Sens. Environ. **74**, 545 (2000)
94. P.C. Smits, IEEE Trans. Geosci. Remote Sens. **40**(4), 801 (2002)
95. J. Pearl, *Probabilistic Reasoning in Intelligent Systems: Networks of Plausible Inference*, 2nd edn. (Morgan Kaufmann, San Francisco, 1988)
96. S. Sarkar, K.L. Boyer, *Computing Perceptual Organization in Computer Vision* (World Scientific, Singapore, 1994)
97. C. Ünsalan, K.L. Boyer, IEEE Trans. Geosci. Remote Sens. **42**(4), 907 (2004)
98. A. Ifarraguerri, C.I. Chang, IEEE Trans. Geosci. Remote Sens. **38**(6), 2529 (2000)
99. A. Hyvärinen, E. Oja, *Independent Component Analysis* (Wiley, New York, 2001)

# Chapter 6
# Land Use Classification using Structural Features

After a thorough survey on land use classification, we begin with a set of measures based on straight lines for Ikonos images in this chapter. Subsequent detailed analyses (counting and classifying dwellings, for example) can then be confined to developed areas in the following chapters. Straight line structures will be more prevalent and more organized in developed areas than in wilderness or rural areas. However, for our measures we only need this assumption to hold locally. Since most buildings are rectangular (or formed of rectangular blocks), and as we observe their layout locally (in $400 \times 400$ $m^2$ windows) we expect this assumption to hold even if it is not satisfied for the overall image (and for most cases it does not). Four of our most promising measures (based on length and contrast) do not depend heavily on this assumption. On the other hand, our remaining measures (orientation, line spacing, and periodicity) depend on this assumption heavily. As expected, this later group could not perform as well as the length and contrast measures experimentally.[1]

Our objective at this stage is the (rough) classification of the image into regions of little or no development (wilderness or rural) and developed regions (urban or residential). We applied Bayes, Parzen window, and nearest neighbor (NN) classifiers to label each image region. Initially, we defined a two-class problem to discriminate "urban" and "not urban" regions and obtained excellent results (roughly 87% correct classification).

Although there has been extensive work on land use classification, no structural approaches to this problem have been reported. Our approach, being totally based on straight lines, offers the first such solution, to our knowledge. This approach shows very promising results in extensive testing over a wide variety of land development patterns. We extract straight lines via line support regions as explained next.

---

[1]The figures in this chapter are obtained from our previous work [1]. Here, they appear with the kind permission of IEEE.

C. Ünsalan, K.L. Boyer, *Multispectral Satellite Image Understanding*,
Advances in Computer Vision and Pattern Recognition,
DOI 10.1007/978-0-85729-667-2_6, 

## 6.1 Line Support Regions (LSR) and Straight Line Extraction

Burns et al. [2] introduced the concept of line support regions, constructed by grouping contiguous pixels of consistent gradient orientation. His gradient calculation used simple $2 \times 2$ masks, which creates many false alarms in our high resolution data. Therefore, we turned to the scale-controllable IIR edge detection filter proposed by Sarkar and Boyer [3] to compute an optimally smoothed gradient. Then, instead of marking edges in the usual way, we organized the image into line support regions in a manner reminiscent of Burns' work.

We calculated the scale-controllable gradients for the $x$ and $y$ components, $G_x$ and $G_y$, by convolving the image in the $x$ and $y$ directions with the Sarkar–Boyer filter. This is achieved as follows. Two non-causal filters, the projection filter $h(n)$ and the edge detection filter $e(n)$, are successively applied to rows and columns of the image. Impulse responses for these filters for $n > 0$ are given in (6.1) and (6.2):

$$h(n) = -e^{-\alpha n} \cos(\beta \alpha n + \pi/2), \tag{6.1}$$

$$e(n) = \frac{e^{-\alpha n}}{\alpha} \left( -\frac{\cos(\beta \alpha n + \pi/2) - \beta \sin(\alpha \beta n + \pi/2)}{1 + \beta^2} \right), \tag{6.2}$$

where $n$ is either $x$ or $y$, depending on orientation. The derivation of these impulse responses is beyond the scope of this book; see [3].

The projection filter smoothes the image to lower the noise. The edge detection filter is designed to respond optimally to step edges in the image. In effect, two (separable) 2D impulse responses are created. The first consists of the projection filter in the $y$ direction and the detection filter in $x$; the second is orthogonal to the first. The projection filter is the integral of the edge detection filter, which greatly simplifies implementation. We used the IIR implementation of these filters as suggested by Sarkar and Boyer.

After extensive testing, we selected a fairly small scale parameter ($\alpha = 1$) to resolve buildings in close proximity. The scale of the filter is directly related to the resolution of the image. Since we are working on satellite images, the scale parameter should be adjusted in accordance with the size of ground features of interest (in pixels) and their spacing. A small scale parameter resolves small and closely located features. On the other hand, a large scale parameter captures general characteristics of the image. This adjustment capability (absent in Burns' implementation) allows users to apply this method to a wide variety of image resolutions. Therefore, our system responds to changes to this parameter. Similarly, we set $\beta = 70$ for implementation. Our system is less sensitive to this parameter, which controls a tradeoff among localization, signal-to-noise ratio out of the filter, and spurious responses (see [3] for details). The impulse responses of projection and edge detection filters with these parameters are given in Fig. 6.1.

Let $f(x, y)$ be the image and let $G_x(x, y) = \{e(x)h(y)\} * f(x, y)$ be the $x$ component of the smoothed gradient, computed by convolving the image with the projection filter (in $y$) and the detection filter (in $x$). Similarly, let $G_y(x, y) =$

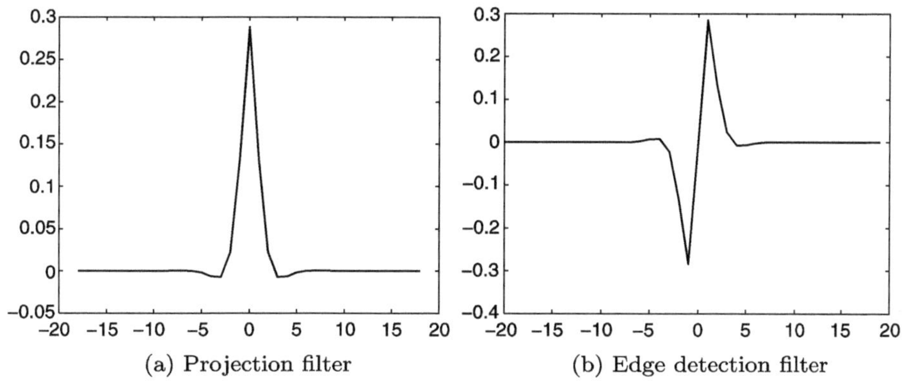


**Fig. 6.1** Impulse responses for the projection and edge detection filters

$\{e(y)h(x)\} * f(x, y)$ be the $y$ component. Then, the smoothed gradient direction at $(x, y)$ is:

$$G(x, y) = \arctan\left(\frac{G_x(x, y)}{G_y(x, y)}\right). \tag{6.3}$$

We discard pixels having gradient magnitude smaller than a given threshold (10 here) to suppress noise. Following Burns et al., we then used two quantizers and a voting method to obtain longer and more consistent line support regions. Each quantizer has *eight* bins of angular resolution $\pi/4$ radians, the second one rotated by $\pi/8$ radians. Two quantizers with shifted bin centers can accommodate candidate line support regions that would be broken in a single quantizer due to minor perturbations.

Burns et al. extracted the lines by intersecting a horizontal plane with the best fit plane to the local image surface. Following Tan [4], we improve on this approach by fitting an ellipse to each line support region perimeter using a Fourier descriptor approximation. The long axis of the ellipse (which requires only the first order descriptors) defines the straight line. This method is faster and more robust than plane fitting. We obtain the best fitting ellipse as follows. A complex periodic function, $u(k) = u(k + rT)$, for any integer values of $k$ and $r$ represents the outer boundary of the line support region. $T$ is the total number of points in the contour. Thus, the complex periodic contour can be approximated by a Fourier series as [5]:

$$\hat{u}(k) = \sum_{n=0}^{T-1} U_n e^{j\frac{2\pi nk}{T}}, \tag{6.4}$$

where $j = \sqrt{-1}$ and

$$U_n = \frac{1}{T}\sum_{k=0}^{T-1} u(k) e^{-j\frac{2\pi nk}{T}}. \tag{6.5}$$

If we insert the complex representation $u(k) = x(k) + jy(k)$ into (6.5), then

$$U_n = \frac{1}{T}\sum_{k=0}^{T-1} x(k)e^{-j\frac{2\pi nk}{T}} + j\frac{1}{T}\sum_{k=0}^{T-1} y(k)e^{-j\frac{2\pi nk}{T}}. \tag{6.6}$$

Applying Euler's formula, we get

$$U_n = \frac{1}{T}\sum_{k=0}^{T-1} x(k)\left(\cos\left(\frac{2\pi nk}{T}\right) - j\sin\left(\frac{2\pi nk}{T}\right)\right)$$
$$+ j\frac{1}{T}\sum_{k=0}^{T-1} y(k)\left(\cos\left(\frac{2\pi nk}{T}\right) - j\sin\left(\frac{2\pi nk}{T}\right)\right), \tag{6.7}$$

$$U_n = \alpha_n + j\beta_n, \tag{6.8}$$

with

$$\alpha_n = \frac{1}{T}\left(\sum_{k=0}^{T-1} x(k)\cos\left(\frac{2\pi nk}{T}\right) + \sum_{k=0}^{T-1} y(k)\sin\left(\frac{2\pi nk}{T}\right)\right), \tag{6.9}$$

$$\beta_n = \frac{1}{T}\left(\sum_{k=0}^{T-1} y(k)\cos\left(\frac{2\pi nk}{T}\right) - \sum_{k=0}^{T-1} x(k)\sin\left(\frac{2\pi nk}{T}\right)\right). \tag{6.10}$$

For our statistical features defined below, we need the center of mass $\mu_{xy}$, the length $l$, and the orientation $o$ of each straight line. Fortunately, from just three Fourier coefficients, $(\alpha_{-1}, \beta_{-1})$, $(\alpha_0, \beta_0)$, and $(\alpha_1, \beta_1)$, we can obtain these quantities from (6.11), (6.12), and (6.13), respectively:

$$\mu_{xy} = (\alpha_0, \beta_0), \tag{6.11}$$

$$l = 2\left(\sqrt{\alpha_1^2 + \beta_1^2} + \sqrt{\alpha_{-1}^2 + \beta_{-1}^2}\right), \tag{6.12}$$

$$o = \frac{1}{2}\left(\arctan(\beta_1/\alpha_1) + \arctan(\beta_{-1}/\alpha_{-1})\right). \tag{6.13}$$

Three sample images and the lines extracted from them for rural, residential, and urban areas are given in Fig. 6.2. The relative degree of organization over the three images is evident in their respective line structures.

## 6.2 Statistical Feature Extraction

We extract statistical features over 400 × 400 pixel windows, with 50% overlap in each direction. This degree of overlap represents a compromise in the interest of computational effort. Therefore, an organization lying partly in one window will

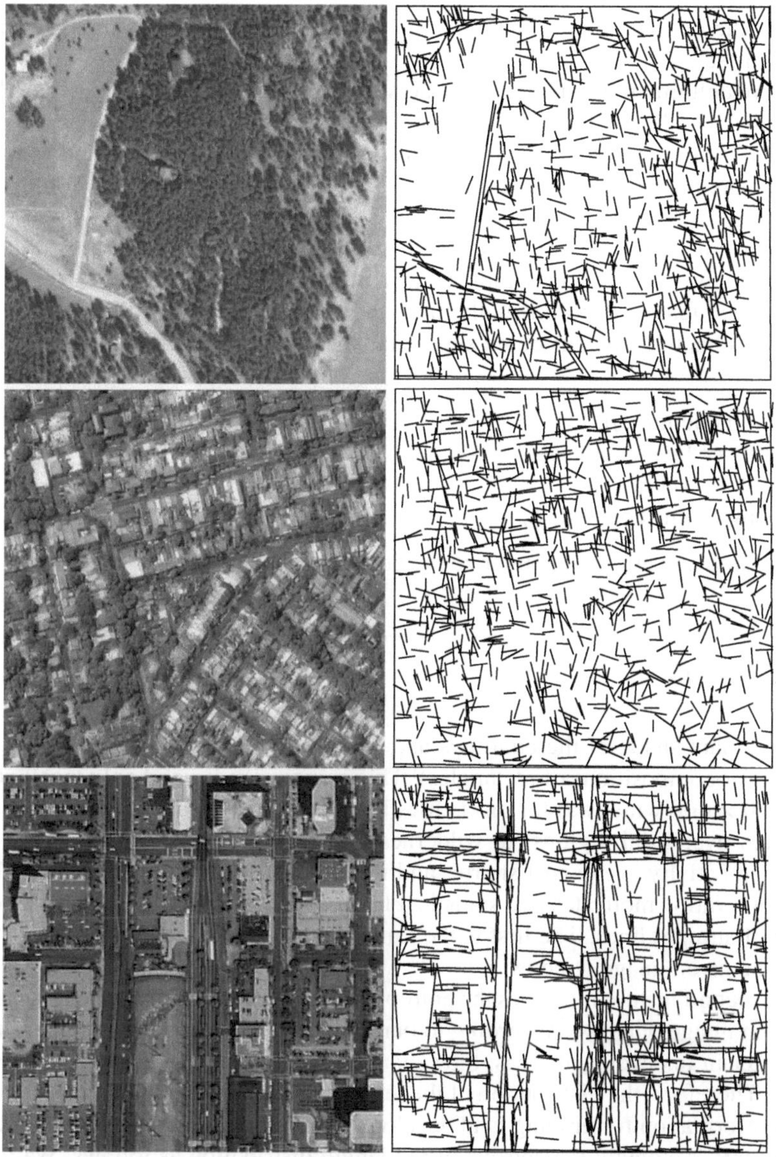

**Fig. 6.2** *First column* (from top to down): rural, residential, and urban regions. *Second column*: lines extracted from them

likely lie totally inside a neighboring window. To evaluate performances of the features, we extracted the ground truths accordingly.

If a line extends beyond the window, that part is truncated and does not enter the calculations for this window. We discard straight lines shorter than five pixels (five meters); such objects can not be approximated robustly due to low signal-to-noise ratios in the contour approximation and certainly do not offer strong evidence for "straightness".

Before defining the statistical measures, we give two definitions that we will use repeatedly. The first is the sample mean $\bar{\mu}$; we assume this is self-explanatory. The second is the entropy $E$ of an approximate probability mass function (pmf). Let the approximate pmf be given as $h(i)$ for $i = 1, \ldots, N$. We calculate the entropy based on (3.13).

For each feature we extract, we obtain an approximate probability mass function from which we compute the sample mean and the entropy. We set the histogram bin size for each feature experimentally; they are not especially critical.

In this chapter, we considered seven statistical features. The most promising of these were:

- Mean line length, $\bar{\mu}_l$
- Entropy of line length, $E_l$
- Mean line contrast, $\bar{\mu}_c$
- Entropy of line contrast, $E_c$.

For completeness, we summarize the other features in Sect. 6.5.

### *6.2.1 Length*

The dominant shape of the buildings and street segments in urban regions is rectangular (or compositions of rectangles). This assumption is valid only locally (such as in $400 \times 400$ windows) for most cities but this suffices because we calculate our measures over local regions. In wilderness (or rural) regions, due to the absence of human activity, we expect shorter and more randomly distributed straight lines. This observation motivates two statistical features to grade land development by the length of straight lines. The first is the sample mean of the length of straight lines, $\bar{\mu}_l$, in the given image window.

The second feature measures the entropy in the distribution of line lengths, $E_l$, in the given image window. According to our hypothesis, not only will image features be shorter in undeveloped areas, they should also be more randomly distributed in length. Therefore, the distribution of the lengths in a given window can also provide useful information for grading land development and we capture the randomness in the entropy. We obtain the entropy of the length distribution as follows. We form a histogram of 37 bins (with $binwidth = 4$ pixels), where the first bin is centered at 5 pixels and the last at 150 pixels. The maximum bin value represents the longest line segment such that the corresponding bin has significant probability and was obtained experimentally. The binwidth is picked to obtain a fairly smooth histogram. These values are not critical in feature extraction. Therefore, our system has a robust characteristics to their small variations.

The histogram is normalized to approximate a probability mass function. Let the normalized vote for bin $i$ be $h_l(i)$ for $i = 1, \ldots, 37$. The entropy of line length is then obtained from (3.13).

### 6.2.2 Contrast

Contrast provides another indicator of the level of development in a given area. In urban regions, for example, we encounter greater variation in the color of buildings, sharp dark-light transitions between asphalt roads and shiny roofs, and the juxtaposition of buildings and vegetation. Such high contrast variations are less likely in wilderness (or rural) regions. This observation lets us define two features based on contrast.

We define the contrast for each line support region from the gradient calculations as follows. Let pixels in a given line support region form a set *LS*; $(x, y) \in LS$ represents a pixel in that line support region. The contrast value for the line support region, *clsr*, is the maximum directional derivative magnitude over the region as given in (6.14):

$$clsr = \max_{(x,y)\in LS}\Big(\max_{d\in\{x,y\}} |G_d(x, y)|\Big). \tag{6.14}$$

Our tests of the true directional derivative magnitude produced similar classification performances; for simplicity, we confine our attention to the $x$ and $y$ directions.

We also tested the average directional derivative magnitude, but the classification results based on this definition were not encouraging because when averaging, low contrast values dominate. The maximum directional derivative definition is biased towards high contrast values, which helps in classification. We obtain our first contrast-based feature, the sample mean $\mu_c$, from *clsr* values.

Our second feature is the line contrast entropy $E_c$. To compute it, we quantize the contrast range to 31 bins (with $binwidth = 95$), where the first bin is centered at 5 and the last bin at 3000. These values are obtained experimentally as in length histograms, and found not to be critical. To build the histogram, each line support region votes with a weight proportional to the length of the line it represents. We then normalize the histogram as usual. Let the normalized vote for bin $i$ be $h_c(i)$ for $i = 1, \ldots, 31$. Then the contrast entropy is obtained from (3.13).

## 6.3 Experimental Results

We applied several classification approaches to both the urban–rural two-class problem and the direct detection of suburban areas using the features described, as well as spatial coherence. We implemented Bayes, Parzen window, and nearest neighbor (NN) classifiers [6]. To focus on the most promising classification approach, we consider only the Bayes classifier in this section. For complete results for all classifiers in different feature spaces, see Sect. 6.5.

### 6.3.1 Dataset and Feature Space

Our data set consists of 28 one-meter Ikonos panchromatic images, 18 drawn from across North America, 10 from different locations around the world. Table 6.1 summarizes the images used in this chapter; they encompass a wide range of cultures and development patterns.

An exhaustive search over the full feature set (all singles, pairs, triples, etc. as in Sect. 6.5) reveals that the best performance is obtained in the $(\bar{\mu}_c - E_c)$ feature space for the Bayes classifier. We used 48 rural and 48 urban image windows selected at random from the overall data set to train each classifier. We tested 2014 new region samples to ascertain the performance of the classifier. A *caveat* on ground truth:

**Table 6.1** Test images

| Geographic location | Image size |
|---|---|
| Colorado, USA | $2002 \times 2002$ |
| Dubai, UAE | $763 \times 953$ |
| California, USA | $640 \times 799$ |
| Colorado, USA | $4177 \times 3733$ |
| New Mexico, USA | $3996 \times 6006$ |
| California, USA | $2001 \times 2001$ |
| Sydney, Australia | $2003 \times 2003$ |
| Colorado, USA | $641 \times 801$ |
| Mississippi, USA | $1992 \times 1995$ |
| Maryland, USA | $7849 \times 3107$ |
| New York, USA | $2000 \times 1997$ |
| Rome, Italy | $1993 \times 1993$ |
| Sanaa, Yemen | $1841 \times 2109$ |
| California, USA | $1992 \times 1996$ |
| Taipei, Taiwan | $2101 \times 2101$ |
| Tokyo, Japan | $1989 \times 2010$ |
| Guayaquil, Ecuador | $1999 \times 1999$ |
| California, USA | $2000 \times 2000$ |
| Cairo, Egypt | $2104 \times 2104$ |
| London, UK | $2000 \times 2000$ |
| California, USA | $2103 \times 2110$ |
| Sydney, Australia | $1430 \times 1539$ |
| Arizona, USA | $1671 \times 1062$ |
| Colorado, USA | $1121 \times 1112$ |
| Arizona, USA | $2007 \times 2007$ |
| California, USA | $1093 \times 1158$ |
| California, USA | $1156 \times 1160$ |
| North Carolina, USA | $1100 \times 772$ |

Some windows straddle transitions from developed to undeveloped areas, and some others encompass lightly developed regions, parkland, and other types of terrain that are hard to classify, even manually.

### 6.3.2 Two-Class Results

With the given feature space and the Bayes classifier, we obtain a 92.0% correct classification for the rural regions over 905 test samples. Similarly, we obtain a 83.0% correct classification for the urban samples over 1109 test samples. The overall classification performance is **87.0%**.

In an attempt to exploit spatial coherence in the two-class case, we applied probabilistic relaxation [7, 8] (to be explained fully in the following chapter). We obtain a 91.2% and 86% correct classification rates for the rural and urban scenes, with a maximum of **88.3%** overall correct classification percentage. Introducing spatial information via probabilistic relaxation improved the correct detection rates slightly; the overall improvement is just 1.3%. More importantly, however, probabilistic relaxation results in more acceptable errors in classification by reducing misses of the urban class.

### 6.3.3 Capabilities and Limitations

To test the limitations of our assumptions regarding the impact of human activity on the emergence of straight lines, we compared results for two highly different regions: downtown San Diego, California and the outskirts of Cairo, Egypt (Fig. 6.3). These two images have almost the same size and the same number of windows (64 in each). This test uses Bayes classification on decidedly the best feature space ($\bar{\mu}_c$–$E_c$) and makes no use of relaxation or feature based grouping.

For the California image, almost all windows are urban, the remainder are shoreline. The correct classification rate for this image is 98.4%. One "rural" region was classified as urban, but it straddles the shoreline. Therefore, even the "ground truth" classification for this window is somewhat debatable.

On the other hand, the Egypt image differs from most of our remaining images. The residential regions in this image are not exactly suburbia. They are not as regular as North American residential regions. The rural regions in this image are formed of sand (unlike any of our training data). For this image, we have a 59.4% correct classification rate. Over 25 rural regions, three of them are labeled as urban. The main reason for this is the region in the upper left corner of the image. Although, it does not show any residential (or urban) characteristics, it has some regularity. It is clearly a collection of man-made structures, but they are too small to be houses or buildings. Over 39 urban regions, 23 are labeled as rural. The reasons for such a high error rate are as follows. Residential regions in this image are not well-structured as

(a) The California image (b) The Egypt image

**Fig. 6.3** Testing the system on specific images

assumed (in the design phase). Some residential regions (around the pyramids) contain demolished houses (ruins), which further decrease the regularity in the region. In transition regions, we see errors due to their mixed characteristics. Also, three-fourths of our data set come either from North America or "North America like" regions (e.g., Australia), so the Egypt results are not surprising. Besides, our classification rate is still above the chance line even for this image.

These two images provide clues regarding the performance capabilities of our classification system. For well-organized regions, it produces very high classification rates. Transitions in urban–rural regions cause classification errors, due to their mixed characteristics. Finally, semi-organized ("old-world") residential regions are more difficult to classify.

## 6.4 Summary of the Classification System

In the previous sections, we discussed each step separately, together with their parameter values. Here, we tabulate the parameters with respect to their nominal values and relative influence on the final performance (sensitivity) in Table 6.2.

Although there are eight parameters in the system, only the value of the gradient scale parameter $\alpha$ has a major impact. If the lines cannot be extracted effectively, all remaining calculations are compromised. The binsize in pixel grouping (in line extraction step) has a medium effect through its influence on the lines extracted. Binsizes in the feature extraction step have only a weak impact on the final classification result because statistical measures summarize these histograms as a whole; no individual entry is critical.

**Table 6.2** Summary of the system parameters

| Step | Section | Parameter | Value | Influence |
|---|---|---|---|---|
| Line extraction | gradient calculation | $\alpha$ (scale) | 1 | high |
| Line extraction | gradient calculation | $\beta$ (spurious response) | 70 | low |
| Line extraction | pixel grouping | binsize | $\pi/4$ | medium |
| Feature extraction | length | binsize | 37 | low |
| Feature extraction | contrast | binsize | 95 | low |
| Classification | Bayes classifier | training set | 96 | medium |

## 6.5 Additional Results

For completeness, we now present the full range of features and feature spaces considered, as well as other classifier designs.

### *6.5.1 Additional Features*

We considered three features beyond those of Sect. 6.2. These are less powerful than the length and contrast based features and require more computation. Their primary limitation is that they depend heavily on city models that do not necessarily hold. However, they do provide reasonable classification performance and may be more useful in specific domains.

#### 6.5.1.1 Orientation

The orientation feature $R_o$ is the cross-correlation between the histogram of straight line orientations in a given image window with an idealized bimodal density function model for local urban areas, having two Gaussian modes separated by $\pi/2$ radians. Since the feature is calculated over a window, this assumption need not to hold for the overall image as is in the previous features. This idealized model works well as long as there is no single dominant mode in street direction in the image window. We build a weighted (based on line length) orientation histogram. We use weighted voting because lines obtained from small line support regions do not have reliable orientations. The orientation range is quantized to 32 bins (with $binwidth = 0.03\pi$ radians), where the first bin is centered at $-\pi/2$ radians and the last bin at $\pi/2$ radians. As in the length and contrast features, the bin size is not crucial.

Let the normalized vote for bin $i$ be $h_o(i)$ for $i = 1, \dots, 32$. We cross-correlate the normalized orientation histogram with the bimodal density function model $g(t)$:

$$g(t) = \frac{1}{2\sqrt{0.2\pi}}\left(e^{\frac{-1}{2}(\frac{t+\pi/4}{0.1})^2} + e^{\frac{-1}{2}(\frac{t-\pi/4}{0.1})^2}\right). \tag{6.15}$$

The denominator of 0.1 in the bimodal density function (actually the standard deviation for each unimodal part) provides a tolerance for the parallelness of straight lines in the image window. This value is obtained from the layout of the buildings in images.

The cross-correlation value given in (6.16) indicates the similarity of the image window to the idealized urban region model while the shift $t$ at which the maximum correlation occurs indicates the dominant street grid orientation. This could prove useful in subsequent analyses.

$$R_o = \max_t \{g(t) \otimes h_o(t)\}. \tag{6.16}$$

#### 6.5.1.2 Line Spacing

This feature captures the spatial distribution of lines in the image window. We represent each line by its center of mass and obtain an entropy measure from the distribution of line spacings. For each line, we locate the centers of mass of the nearest two lines of similar length ($\pm 3$ pixels). The spacing between the original line and each neighbor is taken to be the Euclidean distance between their centers of mass. We repeat this calculation for all lines in the given window.

We compute the histogram of line spacings over 70 bins (with $binwidth = 2$ pixels), with the first bin centered at 2 pixels and the last at 140 pixels. Let the normalized vote for bin $i$ be $h_{dcom}(i)$ for $i = 1, \ldots, 70$. The entropy of line spacing, $E_s$, is then obtained from (3.13).

#### 6.5.1.3 Periodicity

Most urban regions (since the early twentieth century) are constructed in accordance with a (set of) zoning plan(s). Therefore, we expect more grid-like structures (such as blocks of buildings) and more periodic substructures in urban regions. This assumption is often violated for older cities, especially outside North America. Therefore, this feature is not a strong one, in general. However, if one's attention is confined to North America (or Australia) where land development continues apace, this feature may prove useful.

Straight lines extracted from well-organized structures should exhibit some form of periodicity in their placement. We infer periodicity through the projection of the line segments onto each of four axes: $y = 0$, $y = x$, $x = 0$, and $y = -x$; the origin is at the center of the window. To make the discussion concrete, let us consider the case for the $x = 0$ (that is, the $y$) axis. The intersection point of each straight line segment (or its extension) with the line $x = 0$ is identified. A function is defined along this axis whose value at each line segment intersection point is the number of pixels in the line segment(s) projecting to that point. Non-intersection points (points not projected to by any line segment in the window) are assigned the value 0. Figure 6.4 illustrates the procedure for the case of three line segments, one of length

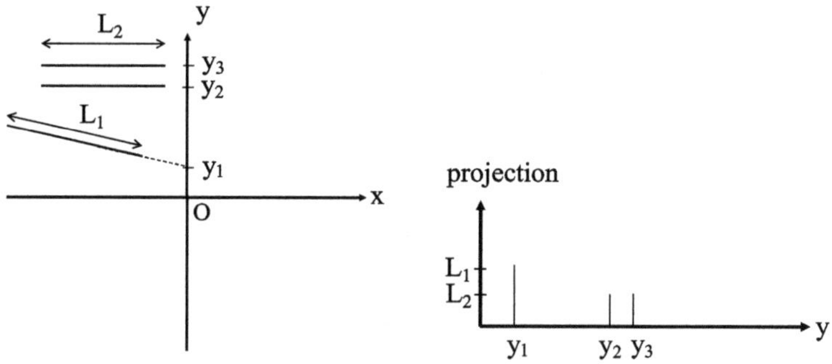


(a) Parallel lines and x-y axes (b) Projection of parallel lines on y axis

**Fig. 6.4** An example of the projection of lines

$L_1$ projecting to location $y = y_1$ on the $x = 0$ axis, and two each of length $L_2$ projecting to $y = y_2$ and $y = y_3$, respectively.

The resulting set of four projection functions are then analyzed for periodicity. Because projections along directions highly oblique to the axis are unreliable, we consider four axes (as mentioned above) and limit the set of projecting line segments to be those oriented within $\pm 45°$ of the direction orthogonal to the current projection axis.

We use the periodicity transform introduced by Sethares and Staley [9] to measure the periodicity of the projection function. They project the given one-dimensional signal onto non-orthogonal periodic subspaces using an M-best periodic decomposition algorithm. We find that the transform coefficient for the first periodic subspace sufficiently represents the periodicity in the projection function. The maximum periodicity $P$ among the four projection functions is taken as the periodicity of the image window.

### *6.5.2 Other Feature Spaces and Classifiers*

Now, we summarize two additional sets of findings with respect to classification. First, we consider the Bayes, Parzen window, and nearest neighbor (NN) classifiers. Then, we compare their performances on the overall feature space. These results will support our decision to use the Bayes classifier with the feature space ($\bar{\mu}_c$–$E_c$) in Sect. 6.3.

We start by tabulating the best classification performances of each classifier in different spaces in Tables 6.3, 6.4, and 6.5. In these tables, each row gives the best set of $n$ features (the best n–D feature space) for $n \in [1, 7]$.

We see that there is a different best feature space for each classifier, but the performances of the classifiers are not statistically different from one another. However, the computational costs of the classifiers, and of the best features for each, do vary.

**Table 6.3** Performances for best feature spaces for the Bayes classifier

| Feature combination | Performance |
|---|---|
| $E_c$ | 86.0 |
| $E_c, \bar{\mu}_c,$ | **87.0** |
| $E_c, \bar{\mu}_c, R_o$ | 86.6 |
| $E_c, \bar{\mu}_c, R_o, E_s$ | 86.5 |
| $E_c, \bar{\mu}_c, R_o, E_s, E_l,$ | 86.6 |
| $E_c, \bar{\mu}_c, R_o, E_s, E_l, P,$ | 86.6 |
| $E_c, \bar{\mu}_c, R_o, E_s, E_l, P, \bar{\mu}_l$ | 86.4 |

**Table 6.4** Performances for best feature spaces for the Parzen window classifier

| Feature combination | Performance |
|---|---|
| $E_c$ | 87.1 |
| $E_c, E_l$ | **87.3** |
| $\bar{\mu}_c, E_s, P$ | 83.5 |
| $E_c, \bar{\mu}_c, E_l, \bar{\mu}_l$ | 75.4 |
| $E_c, \bar{\mu}_c, E_l, \bar{\mu}_l, E_s$ | 81.5 |
| $E_c, \bar{\mu}_c, E_l, E_s, P, R_o$ | 82.8 |
| $E_c, \bar{\mu}_c, E_l, \bar{\mu}_l, E_s, P, R_o$ | 82.3 |

**Table 6.5** Performances for best feature spaces for the NN classifier

| Feature combination | Performance |
|---|---|
| $E_c$ | 87.6 |
| $E_c, E_l$ | **87.8** |
| $E_c, E_s, R_o$ | 84.5 |
| $E_l, E_s, R_o, \bar{\mu}_c$ | 84.0 |
| $E_c, E_l, R_o, \bar{\mu}_c, P$ | 83.2 |
| $E_c, E_l, R_o, \bar{\mu}_c, P, E_s$ | 83.2 |
| $E_c, E_l, R_o, \bar{\mu}_c, P, E_s, \bar{\mu}_l$ | 82.8 |

We can use computational cost and performance together to select the best feature space–classifier combination.

Next, we tabulate the classification performances for all 1D and 2D feature spaces for each classifier in Tables 6.6, 6.7, and 6.8. In these tables, diagonal cells represent the classification performance of each feature alone, off-diagonal cells represent the classification performance of the corresponding 2D feature space.

These tables show that the length and contrast based features are highly effective for classification. However, although they perform above the chance line, the rotation and projection features perform poorly relative to the others.

Finally, we provide the classification performance (rural, urban, and overall) as well as the best feature space for each classifier separately in Table 6.9.

**Table 6.6** The Bayes classification results in percentages for all 1D and 2D feature spaces

| | $\bar{\mu}_c$ | $E_c$ | $\bar{\mu}_l$ | $E_l$ | $E_s$ | $P$ | $R_o$ |
|---|---|---|---|---|---|---|---|
| $\bar{\mu}_c$ | 85.8 | **87.0** | 83.9 | 85.9 | 79.7 | 80.5 | 82.4 |
| $E_c$ | | 86.0 | 84.0 | 86.0 | 79.1 | 77.9 | 84.5 |
| $\bar{\mu}_l$ | | | 82.7 | 83.9 | 83.5 | 83.3 | 84.1 |
| $E_l$ | | | | 86.0 | 84.5 | 86.4 | 84.9 |
| $E_s$ | | | | | 72.4 | 77.1 | 73.3 |
| $P$ | | | | | | 68.2 | 68.6 |
| $R_o$ | | | | | | | 67.8 |

**Table 6.7** The Parzen window classification results in percentages for all 1D and 2D feature spaces

| | $\bar{\mu}_c$ | $E_c$ | $\bar{\mu}_l$ | $E_l$ | $E_s$ | $P$ | $R_o$ |
|---|---|---|---|---|---|---|---|
| $\bar{\mu}_c$ | 84.7 | 86.8 | 83.4 | 85.5 | 81.2 | 81.5 | 78.3 |
| $E_c$ | | 87.1 | 85.4 | **87.3** | 83.2 | 82.3 | 83.9 |
| $\bar{\mu}_l$ | | | 78.3 | 81.6 | 79.7 | 79.2 | 72.8 |
| $E_l$ | | | | 82.7 | 83.5 | 84.2 | 76.5 |
| $E_s$ | | | | | 72.9 | 73.3 | 72.6 |
| $P$ | | | | | | 65.9 | 62.5 |
| $R_o$ | | | | | | | 58.8 |

**Table 6.8** The NN classification results in percentages for all 1D and 2D feature spaces

| | $\bar{\mu}_c$ | $E_c$ | $\bar{\mu}_l$ | $E_l$ | $E_s$ | $P$ | $R_o$ |
|---|---|---|---|---|---|---|---|
| $\bar{\mu}_c$ | 85.8 | 85.9 | 85.9 | 85.9 | 85.9 | 85.9 | 85.9 |
| $E_c$ | | 87.6 | 84.4 | **87.8** | 87.1 | 76.2 | 88.0 |
| $\bar{\mu}_l$ | | | 81.6 | 81.7 | 82.6 | 79.6 | 81.4 |
| $E_l$ | | | | 85.0 | 83.7 | 71.4 | 83.6 |
| $E_s$ | | | | | 72.4 | 68.7 | 74.1 |
| $P$ | | | | | | 61.2 | 63.3 |
| $R_o$ | | | | | | | 63.7 |

**Table 6.9** Performance comparison

| Classifier | Best feature space | Performance (%) | | |
|---|---|---|---|---|
| | | Rural | Urban | Average |
| Bayes | $\bar{\mu}_c$–$E_c$ | 92.0 | 83.0 | 87.0 |
| Parzen window | $E_c$–$E_l$ | 89.3 | 85.7 | 87.3 |
| NN | $E_c$–$E_l$ | 91.0 | 85.2 | 87.8 |

Each classifier presents both benefits and shortcomings. The Bayes classifier comes at low computational cost, but assumes a parametric model for the data distribution, nearly always Gaussian. If the data represent such a characteristic, the classifier performs well. If not, it may be possible to model the data as a mixture of multiple Gaussians; we have not pursued this additional complexity in this chapter. The Parzen window and NN classifiers do not assume any particular a priori distribution; they model the data distribution locally. The main drawback of these two methods is their computational cost. Although they may represent non Gaussian distributions more robustly than the Bayes classifier, they did not perform appreciably better on our data. Therefore, we recommend the Bayes classifier in this domain.

## 6.6 Summary of the Chapter

We have presented a new approach to assess the degree of land development. While spectral signatures have been exploited for years in land use classification, our objective was to learn what information regarding land development could be extracted—efficiently—from the photometric *structure* in the image. We based the assessment on the photometric and geometric characteristics of straight line segments. Although edge detection may be used to extract lines, we used a more robust *straight line* extraction method based on regions of consistent gradient orientation.

Among several features we developed, those based on length and contrast proved to be the most promising for classification. The strength of these features is that they do not depend heavily on the city model; they apply to most cities around the world, particularly over 400 $m^2$ windows. The weaker features are those more dependent on a specific city model as described in Sect. 6.5. The model assumes a high(er) degree of organization, which may not hold for many old world cities. However, if one's focus were on urban and rural discrimination in North America (and Australia, for example), then these features may prove more useful.

We tested both parametric (Bayes) and non-parametric (Parzen window and nearest-neighbor) classifiers. Although these classifiers differ in their assumptions with respect to the underlying data distribution, they offer similar performance on our data. However, once trained, the Bayes classifier requires the least computation; the nearest neighbor the most. Therefore, we would recommend the Bayes classifier for a production system using this type of data. With probabilistic relaxation we obtained slightly improved classification performance. However, this improvement is marginal considering its additional computational cost, and we would not recommend it. Next, we will add multispectral information to classify land use.

## 6.7 Problems

**6.1** Why is structural information important for land use classification?

**6.2** What is the most time-consuming part in LSR method?

**6.3** How is LSR different from other line extraction methods?

**6.4** What is the main limitation for structural features in general?

**6.5** Sort the classifiers used in this chapter based on their computation load.

**6.6** Sort the structural features in this chapter based on their computation load.

**6.7** (**Open ended question**) What may be other candidate structure information to be used as a feature?

## References


1. C. Ünsalan, K.L. Boyer, IEEE Trans. Geosci. Remote Sens. **42**(7), 1575 (2004)
2. J.B. Burns, A.R. Hanson, E.M. Riseman, IEEE Trans. Pattern Anal. Mach. Intell. **8**(4), 425 (1986)
3. S. Sarkar, K.L. Boyer, IEEE Trans. Pattern Anal. Mach. Intell. **13**(11), 1154 (1991)
4. S.G. Tan, Image Feature Extraction: Line Detection and Organization. Msc. thesis, The Ohio State University (1990)
5. A.V. Oppenheim, R.W. Schafer, J.R. Buck, *Discrete Time Signal Processing*, 2nd edn. (Prentice Hall, New York, 1999)
6. R.O. Duda, P.E. Hart, D.G. Stork, *Pattern Classification*, 2nd edn. (Wiley Interscience, New York, 2001)
7. A. Rosenfeld, R.A. Hummel, S.W. Zucker, IEEE Trans. Syst. Man Cybern. **6**, 420 (1973)
8. H. Yamamoto, Comput. Vis. Graph. Image Process. **10**, 256 (1979)
9. W.A. Sethares, T.W. Staley, IEEE Trans. Pattern Anal. Mach. Intell. **47**(11), 2953 (1999)

# Chapter 7
# Land Use Classification via Multispectral Information

In Chap. 6, our land use classification approach was based on the organization of straight lines (structure) in panchromatic images. It is well-known that multispectral information also offers a great deal of information for land use classification. This chapter describes an approach to combining structural information, obtained from 1 m panchromatic Ikonos images with spectral information, obtained from the corresponding 4 m multispectral images with application to identifying areas of significant land development. There are several contributions in the literature in which spatial and spectral features have been combined in land use classification and related problems. However, none to date use the line support region structural feature, as we do.[1]

Finally, we introduce additional spatial information, over a broader area than the structural information captured in the line support regions, by means of probabilistic relaxation. Although relaxation improves classification slightly, the improvement comes at substantial computational cost. Therefore, we recommend that this approach be used only in applications where improvement is absolutely necessary.

## 7.1 Introduction

It is well-known that multispectral sensing offers a great deal of information for land use classification; see, for example, [1–10]. These contributions all extract and/or classify individual features (in the cartographic sense) or regions by fusing spectral and spatial information, but do not combine the information into hybrid features (in the statistical pattern recognition sense) as we do. That is, the combination of information from the spectral and spatial domains occurs late in the process in the prior work; we combine the information earlier in the analysis. Also, our spatial (more specifically, structural) features are based on consistent gradient orientation in the

[1] The figures in this chapter are obtained from our previous work [14]. Here, they appear with the kind permission of IEEE.

C. Ünsalan, K.L. Boyer, *Multispectral Satellite Image Understanding*,
Advances in Computer Vision and Pattern Recognition,
DOI 10.1007/978-0-85729-667-2_7, 

panchromatic image in a manner not used in the work cited above. In this chapter, we use *NDVI* and $\theta$ considered in Chap. 3. The multispectral features capture statistical properties of the *NDVI* and $\theta$ measures.

## 7.2 Statistical Feature Extraction

Before defining our statistical measures, let us give the definitions we will use repeatedly. Besides the ones defined in Chap. 6, we have the sample variance $\sigma^2$, the sample skewness $\gamma_1$, and the sample kurtosis $\gamma_2$; we assume these to be self-explanatory.

### 7.2.1 Structural Features

We introduced these features in the previous chapter where we explored a wide range of possibilities. Based on those results, we extract four features based on the statistical properties of the lines in panchromatic images:

- Mean line length, $\mu_l$
- Entropy of line length, $E_l$
- Mean line contrast, $\mu_c$
- Entropy of line contrast, $E_c$.

where the above reference and Sect. 6.1 explain how to extract these features in detail.

### 7.2.2 Multispectral Features

To summarize the multispectral information, we use the *NDVI* and $\theta$. Most wilderness regions have high vegetation density. As we move from wilderness to the city center, we encounter less and less vegetation. The *NDVI* and $\theta$ therefore offer evidence for the level of development.

The top row of Fig. 7.1 presents three sample images from our dataset, one each from urban, residential, and wilderness regions. We give the extracted straight lines from each image in the second row. The relative degree of organization over the three images is evident in their respective line structures. We also give the *NDVI* and $\theta$ representations of these images in the bottom two rows. In these figures, blue corresponds to the lowest *NDVI* value and red corresponds to the highest. Similar to the organization of the lines, we can see a separation among these three regions in the *NDVI* and $\theta$ images according to the degree and spatial distribution of vegetated regions.

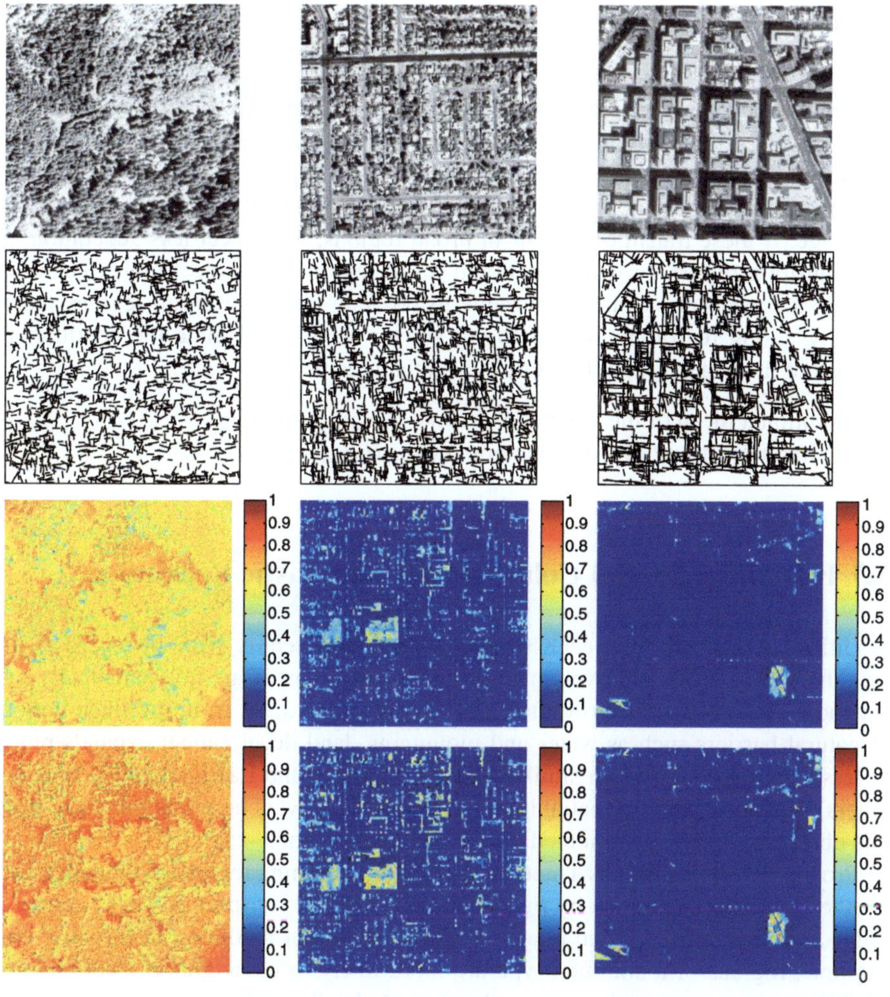

**Fig. 7.1** *First row* (from left to right): Rural, residential, and urban regions. *Second row*: Lines extracted from them. *Third row*: Corresponding *NDVI* images. *Fourth row*: Corresponding $\theta$ images

As multispectral features, we introduce the well-known statistics obtained from the overall vegetation index representation of the image. These are the mean $\mu$, variance $\sigma^2$, skewness $\gamma_1$, and the kurtosis $\gamma_2$ of the index representation. We compute these values both for the *NDVI* and $\theta$ (linearized version of the *NDVI*) separately. The multispectral features we extract are:

- Mean of the vegetation index, $\mu_{ndvi}$ and $\mu_\theta$
- Variance of the vegetation index, $\sigma^2_{ndvi}$ and $\sigma^2_\theta$
- Skewness of the vegetation index, $\gamma_{1,ndvi}$ and $\gamma_{1,\theta}$
- Kurtosis of the vegetation index, $\gamma_{2,ndvi}$ and $\gamma_{2,\theta}$.

### 7.2.3 Hybrid Features

The following two hybrid features combine structural and multispectral information. As the first hybrid feature, we compute the conditional mean of the vegetation index (*NDVI* or $\theta$), given that the index pixels are on a line support region. Our reasoning for this feature is that line support regions are strong indicators of the transitions in an image. If these transitions occur in urban regions, we expect less vegetation in this transition region. In wilderness (and rural) regions, we expect more. This can provide a constraint on our mean value to separate rural and urban regions more robustly. Using reasoning similar to the above, we also calculate the conditional entropy of the vegetation index. The hybrid features we extract are:

- Conditional mean of the vegetation index, $H\mu_{ndvi}$ and $H\mu_\theta$
- Conditional entropy of the vegetation index, $HE_{ndvi}$ and $HE_\theta$.

## 7.3 Exploiting Spatial Coherence: Probabilistic Relaxation

Traditional classifiers treat each window as a distinct entity without regard to its neighbors. Yet, levels of development in neighboring areas are not statistically independent. Cities tend to grow from a core outwards; apart from the intervention of natural barriers such as water and mountains, land development expands more or less coherently. Based on similar reasoning, Kettig and Landgrebe [11] (among others) worked on adding spatial information to land cover classification. To exploit this observation, we implemented probabilistic relaxation.

Probabilistic relaxation is a method that allows adjoining regions to influence one another's label probabilities by means of compatibility constraints. These compatibility constraints encode the relative (in)consistency of adjoining label-pairs. For example, regions of a common type are more likely to be neighbors than regions of different types, especially, say rural–urban. A detailed discussion of probabilistic relaxation labeling is beyond the scope of this book; see [12].

To initialize the relaxation, we use the output of the original classifier. If the Parzen window classifier is in use, we set the initial membership value in a given class equal to the corresponding (relative) weight in the Parzen window. For the Bayesian classifier, we begin with the corresponding likelihood functions $M_u$ (urban) and $M_r$ (rural):

$$M_u = \exp\left[(\mathbf{x} - \bar{\mathbf{x}}_u)\boldsymbol{\Sigma}_u^{-1}(\mathbf{x} - \bar{\mathbf{x}}_u)^T - \ln|\boldsymbol{\Sigma}_u|\right], \tag{7.1}$$

$$M_r = \exp\left[(\mathbf{x} - \bar{\mathbf{x}}_r)\boldsymbol{\Sigma}_r^{-1}(\mathbf{x} - \bar{\mathbf{x}}_r)^T - \ln|\boldsymbol{\Sigma}_r|\right], \tag{7.2}$$

where $\bar{\mathbf{x}}$ is the sample mean and $\boldsymbol{\Sigma}$ the covariance matrix for that class. We obtain the normalized class membership value for each sample by dividing its class membership by the sum of its membership values.

In probabilistic relaxation, the label probabilities for each object are updated based on the conditional probabilities of occurrences of neighboring pairs. The object label probabilities are updated iteratively until a globally consistent set of labels is obtained or (usually) until the label probabilities stabilize. The effect of neighboring objects is encoded through compatibility coefficients, $r_d(\lambda, \lambda')$, where $\lambda$ is a candidate label on the center region and $\lambda'$ is a candidate label on the adjacent region in direction $d$. Since any rural region can be in any direction from any urban region, our compatibility coefficients $r_d(\lambda, \lambda')$ are independent of direction, even though the formalism permits it to be direction dependent. $r_d(\lambda, \lambda')$ is positive for label pairs that are likely (mutually supportive) and negative for those that are mutually inhibitory. We used Yamamoto's method [13] to construct the compatibility coefficients from the prior and conditional label probabilities as follows:

$$r_d(\lambda, \lambda') = \begin{cases} \frac{1}{1-p(\lambda)}(1 - \frac{p(\lambda)}{p_d(\lambda|\lambda')}) & \text{if } p(\lambda)p(\lambda') < p(\lambda, \lambda'), \\ \frac{p_d(\lambda|\lambda')}{p(\lambda)} - 1 & \text{otherwise.} \end{cases} \tag{7.3}$$

We assume equal priors: $p(\lambda) = 1/2$ for $\lambda \in \{\text{urban, rural}\}$ prior to any measurement computation. Label probabilities are then updated in a parallel-iterative fashion as:

$$p_i^{(k+1)}(\lambda) = \frac{p_i(\lambda)[1 + q_i^{(k)}(\lambda)]}{\sum_{\lambda'}\{p_i^{(k)}(\lambda')[1 + q_i^{(k)}(\lambda')]\}}, \tag{7.4}$$

$$q_i^k(\lambda) = \frac{1}{4} \sum_{d \in (N,S,E,W)} \sum_{\lambda'} r_d(\lambda, \lambda') p_d^{(k)}(\lambda'), \tag{7.5}$$

where $p_d^{(k)}(\lambda')$ is the $k$th estimate of the probability of label $\lambda'$ occurring at the adjacent region in the direction $d$. Similar to $r_d$, in our case $p_d$ is the same for all directions. The conditional label-pair probabilities $p_d(\lambda|\lambda')$ were estimated manually from the same training data used earlier. The effect of (7.5) is to compute an adjustment to the probability of label $\lambda$ on region $i$ according to the compatibilities and current estimates of label probabilities on neighboring regions. Equation (7.4) renormalizes the label probabilities after all adjustments have been applied.

We applied the probabilistic relaxation algorithm to the two-class problem as follows. Class label probabilities were taken as normalized class membership values as obtained (initially) from the $\mu_\theta, \sigma_\theta^2, \gamma_{2,\theta}, H\mu_\theta, HE_\theta$ feature space. After exhaustive search over all feature combinations, this gave the best Bayes classifier. These label probabilities were then adjusted by the repeated action of (7.4) and (7.5) until they stabilize. We then applied a *maximum a posteriori* (MAP) decision to the probabilistic relaxation algorithm results for the final labeling.

For demonstration, we first take the Colorado image in Fig. 7.2, which is composed mostly of rural (or wilderness) regions. Fig. 7.3 gives the probabilistic relaxation results for the Colorado image at four stages of the process. In these figures, only the regions labeled as urban are given, the rest is labeled as white. We observe that the region initially labeled as urban shrinks as the relaxation proceeds,

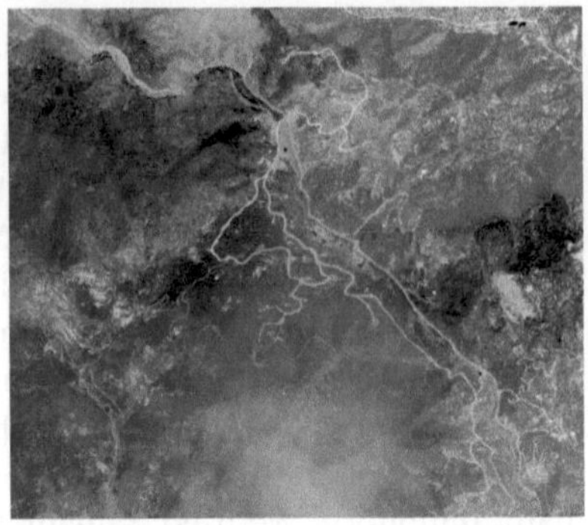

**Fig. 7.2** The Colorado image

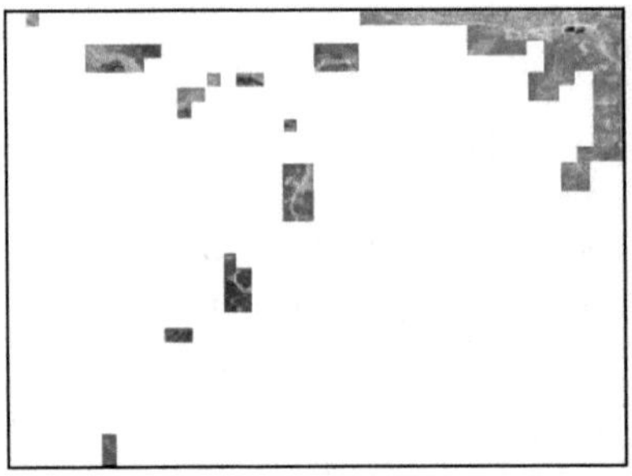

(a) iteration=1

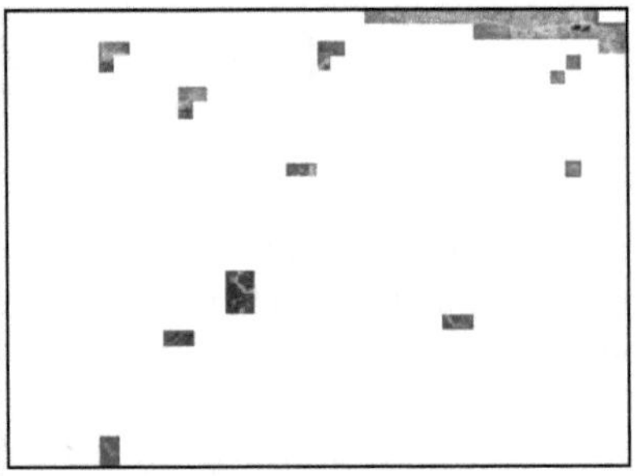

(b) iteration=4

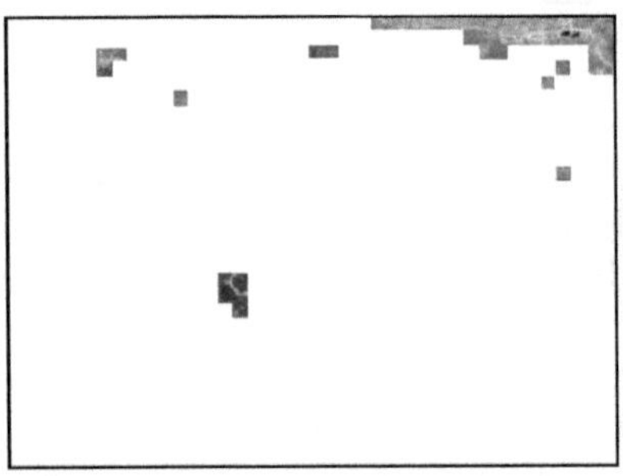

(c) iteration=7

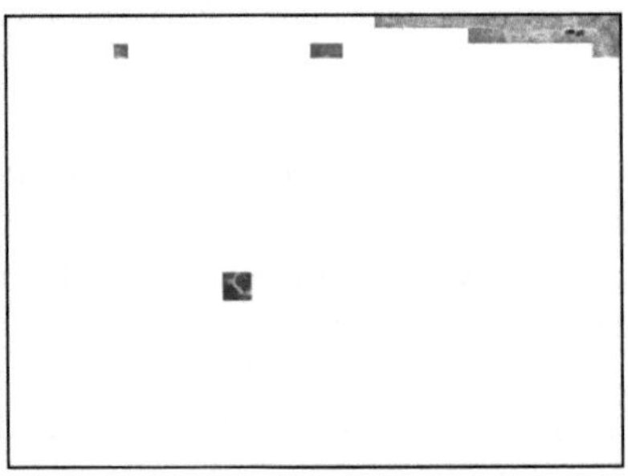

(d) iteration=10

**Fig. 7.3** The Colorado image, probabilistic relaxation iterations

correctly. Since we had pockets of (erroneous) urban labels among a large block of rural (wilderness) regions, and since such an arrangement is punished in the relaxation, the labeling improves. If we compare the initial and the final images, we observe that the urban region false alarms are greatly reduced.

Two more examples of probabilistic relaxation are the Rome image given in Fig. 7.4 and Ecuador image given in Fig. 7.6. For the Rome image, the region ini-

**Fig. 7.4** The Rome image

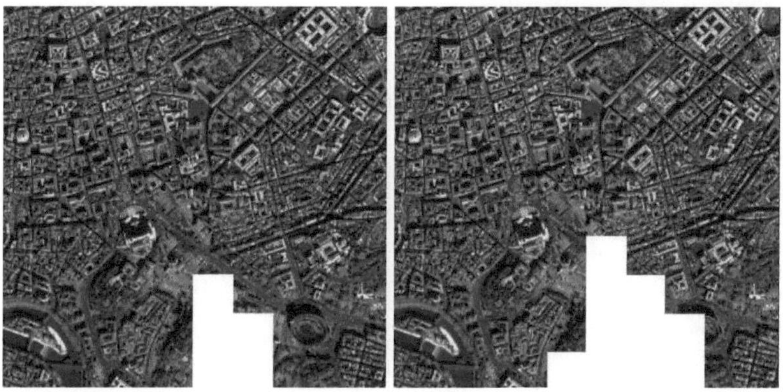

(a) Before probabilistic relaxation

(b) After probabilistic relaxation

**Fig. 7.5** The Rome image, probabilistic relaxation results

tially labeled as rural expands as the relaxation proceeds, given in Fig. 7.5, as desired. Similarly, for the Ecuador image, the region initially labeled as rural shrinks and the region initially labeled as urban expands as the relaxation proceeds, given in Fig. 7.7, as desired.

## 7.4 Experimental Classification Results

This section considers the experiment setup for this chapter. The first subsection deals with the data set used in experiments. In the second subsection, we summarize the classifiers used in the experiments. We tested our features using the standard classifiers (Bayes, Parzen window, and the nearest neighbor (NN)), and a relaxation based MAP decision. We tabulated the classification performances of structural, multispectral, and hybrid features separately so that we could compare them. For

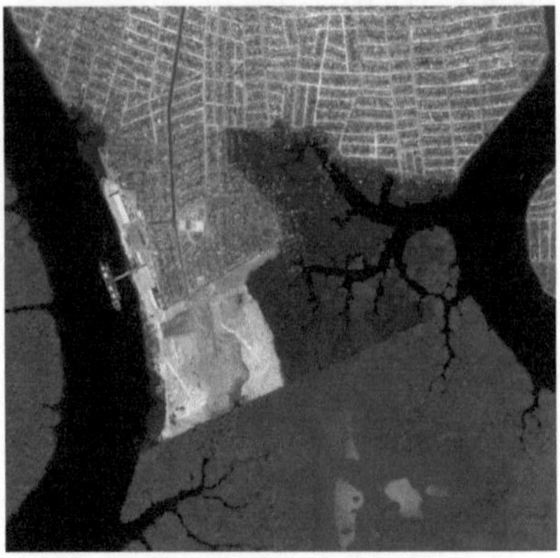

**Fig. 7.6** The Ecuador image

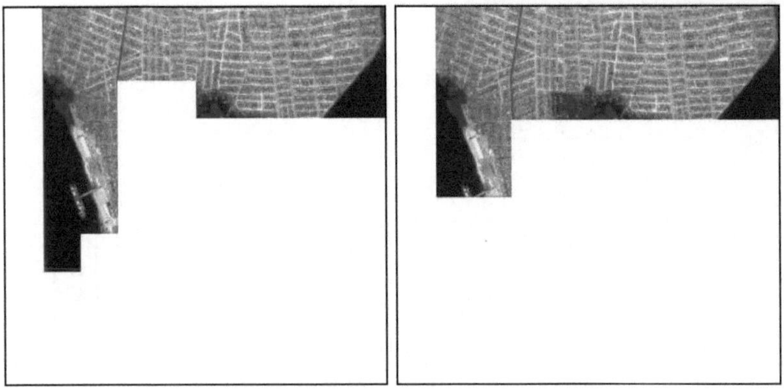

(a) Before probabilistic relaxation

(b) After probabilistic relaxation

**Fig. 7.7** The Ecuador image, probabilistic relaxation results

each set of features (structural, multispectral, and hybrid), we applied an exhaustive search over the full feature set (all singles, pairs, triples, etc.) to obtain the best performing feature space. Based on the classification results obtained, we compared the performance of structural, multispectral, and hybrid features. To observe the effect of vegetation index linearization, we also compared the classification performances of the *NDVI* and $\theta$ based features separately. The last subsection summarizes the sources of error and possible ways to eliminate them.

### *7.4.1 Data Set Specifications*

Our data set is composed of one meter resolution panchromatic and the corresponding four meter resolution multispectral Ikonos satellite images of North American regions. Table 7.1 lists all of the images forming this data set. These test images are labeled by authors. Unfortunately, we were unable to obtain a labeled data set by experts. Therefore, our results may not represent the ideal classification results. However, our labeling of the data set is consistent and true, to the best of our knowledge. As for problems on labeling, some windows straddle transitions from developed to undeveloped areas, and some others encompass lightly developed regions, parkland, and other types of terrain that are hard to classify, even manually.

We extract the features over 800 × 800 meter windows and discard straight lines shorter than five pixels (five meters); such objects cannot be approximated robustly due to low signal-to-noise ratios in the contour approximation and certainly do not offer strong evidence for "straightness". In the following results, each 800 × 800 meter window is considered as a whole for feature extraction.

For the present experiments, we fixed the window size experimentally. However, before deploying a production system, one would want to do a throughout experimental investigation to determine the optimal window size. Fortunately, the results are not highly sensitive to the window size selection, except that the window size must bear some relationship to the size of physical features on the ground, in particular, urban areas. The size of the typical city block and the width of city or residential streets, for instance, entered our "calculations" when considering the issue.

## *7.4.2 Classifier Design*

We give the feature spaces, and number of training and test samples for each classifier separately. Each classifier has benefits and shortcomings.

### 7.4.2.1 Bayes Classifier Specifications

To train the Bayes classifier, we used 112 rural and 136 urban image windows selected at random from the overall data set. We then tested 5657 new region samples (3903 rural, 1754 urban) to ascertain the classification performances. We give the results in Table 7.2 for the best-performing feature space for each feature set.

### 7.4.2.2 Parzen Window Classifier Specifications

To train the Parzen window classifier, we used 419 rural and 419 urban image windows selected at random from the overall data set. We then tested 5160 new region samples (3449 rural, 1611 urban) to ascertain the classification performances. We give the results in Table 7.3 for the best-performing feature space for each feature set.

**Table 7.1** Test images

| Geographic location | Acquisition date | Image size ($m^2$) |
|---|---|---|
| Arizona | 7/23/00 5:55 PM | $6508 \times 6500$ |
| Arizona | 8/14/00 5:57 PM | $5588 \times 4696$ |
| California | 4/5/00 6:16 PM | $11372 \times 8992$ |
| California | 3/15/00 6:48 PM | $22092 \times 13312$ |
| California | 3/18/00 6:59 PM | $19608 \times 15488$ |
| California | 3/18/00 6:59 PM | $11264 \times 15544$ |
| California | 3/18/00 6:59 PM | $10868 \times 15580$ |
| California | 3/18/00 6:59 PM | $23344 \times 13808$ |
| California | 3/18/00 6:59 PM | $12544 \times 13860$ |
| California | 3/18/00 6:59 PM | $12544 \times 13860$ |
| California | 3/26/00 6:51 PM | $13824 \times 12972$ |
| California | 3/26/00 6:51 PM | $13596 \times 12972$ |
| California | 8/18/00 6:40 PM | $16772 \times 12456$ |
| Florida | 5/20/00 3:44 PM | $7424 \times 6760$ |
| Idaho | 5/30/00 6:21 PM | $5188 \times 7176$ |
| Indiana | 5/24/00 4:25 PM | $6436 \times 6448$ |
| Indiana | 5/24/00 4:24 PM | $6492 \times 6284$ |
| Indiana | 7/1/00 4:12 PM | $8160 \times 8140$ |
| Indiana | 6/1/00 4:17 PM | $5100 \times 5092$ |
| Indiana | 9/27/00 4:22 PM | $6436 \times 6448$ |
| Indiana | 8/14/00 4:17 PM | $6436 \times 6448$ |
| Kansas | 5/11/00 4:50 PM | $11436 \times 11196$ |
| Maryland | 6/3/00 3:50 PM | $11348 \times 11132$ |
| Maryland | 4/1/00 3:52 PM | $10000 \times 10000$ |
| Minnesota | 2/16/00 4:47 PM | $5916 \times 8660$ |
| Minnesota | 4/25/00 5:03 PM | $5916 \times 8660$ |
| Mississippi | 7/26/00 4:25 PM | $5316 \times 7008$ |
| Mississippi | 7/21/00 4:43 PM | $6172 \times 4008$ |
| Mississippi | 8/3/00 4:17 PM | $5076 \times 5084$ |
| Mississippi | 9/27/00 4:23 PM | $5076 \times 5084$ |
| Mississippi | 7/21/00 4:43 PM | $5216 \times 4996$ |
| Missouri | 2/29/00 4:25 PM | $7340 \times 7360$ |
| Missouri | 2/29/00 4:25 PM | $7340 \times 7360$ |
| New Mexico | 5/23/00 5:28 PM | $10500 \times 10776$ |
| Oklahoma | 5/22/00 4:52 PM | $11324 \times 11004$ |
| Oregon | 8/8/00 7:14 PM | $11128 \times 11092$ |
| South Dakota | 5/20/00 5:16 PM | $11420 \times 11260$ |
| Wyoming | 6/26/00 6:06 PM | $14068 \times 10188$ |

**Table 7.2** Comparison of performances for the Bayes classifier

| Method | Feature space | Performance (%) | | |
|---|---|---|---|---|
| | | rural | urban | average |
| Structural | $\mu_c, E_c$ | 79.6 | 69.4 | 76.5 |
| Multispectral (*NDVI*) | $\mu_{ndvi}, \sigma^2_{ndvi}$ | 83.0 | 71.6 | 79.4 |
| Multispectral ($\theta$) | $\mu_\theta, \sigma^2_\theta$ | 82.1 | 76.1 | 80.2 |
| Hybrid | $\mu_\theta, \sigma^2_\theta, \gamma_{2,\theta}, H\mu_\theta, HE_\theta$ | 86.1 | 86.2 | 86.1 |
| MAP decision | $\mu_\theta, \sigma^2_\theta, \gamma_{2,\theta}, H\mu_\theta, HE_\theta$ | 89.4 | 84.8 | 88.0 |

**Table 7.3** Comparison of performances for the Parzen window classifier

| Method | Feature space | Performance (%) | | |
|---|---|---|---|---|
| | | rural | urban | average |
| Structural | $\mu_c, E_c$ | 88.7 | 54.6 | 75.3 |
| Multispectral (*NDVI*) | $\mu_{ndvi}, \sigma^2_{ndvi}, \gamma_{1,ndvi}, \gamma_{2,ndvi}$ | 91.7 | 70.2 | 84.7 |
| Multispectral ($\theta$) | $\mu_\theta, \sigma^2_\theta, \gamma_{1,\theta}, \gamma_{2,\theta}$ | 92.2 | 71.4 | 85.5 |
| Hybrid | $\mu_\theta, \sigma^2_\theta, \gamma_{2,\theta}, H\mu_\theta, HE_\theta$ | 92.6 | 72.7 | 86.2 |
| MAP decision | $\mu_\theta, \sigma^2_\theta, \gamma_{2,\theta}, H\mu_\theta, HE_\theta$ | 87.5 | 83.7 | 86.4 |

**Table 7.4** Comparison of performances for the NN classifier

| Method | Feature space | Performance (%) | | |
|---|---|---|---|---|
| | | rural | urban | average |
| Structural | $\mu_c, E_c$ | 86.8 | 56.4 | 76.2 |
| Multispectral (*NDVI*) | $\mu_{ndvi}, \sigma^2_{ndvi}, \gamma_{1,ndvi}$ | 91.0 | 68.3 | 83.6 |
| Multispectral ($\theta$) | $\mu_\theta, \sigma^2_\theta, \gamma_{1,\theta}$ | 91.5 | 69.4 | 84.3 |
| Hybrid | $\mu_\theta, \sigma^2_\theta, \gamma_{2,\theta}, H\mu_\theta, HE_\theta$ | 92.6 | 72.3 | 86.0 |

#### 7.4.2.3 Nearest Neighbor (NN) Classifier Specifications

To train the NN classifier, we used 419 rural and 419 urban image windows selected at random from the overall data set. We then tested 5160 new region samples (3411 rural, 1649 urban) to ascertain the classification performances. We give the results for the best-performing feature space for each feature set in Table 7.4.

### *7.4.3 Comparison of Classification Results*

In this section, we compare classification results from different perspectives. They may give insight about the multispectral and hybrid features.

#### 7.4.3.1 Comparison of Structural, Multispectral, and Hybrid Features

We tabulate the classification performances of structural, multispectral (*NDVI* and $\theta$ separately), hybrid features (with additional multispectral features), and MAP decisions (for Bayes and Parzen window) in Fig. 7.8. In these figures, we provide the average classification performances and urban classification performances separately. By average classification we mean the correct classification rates of urban and rural test samples together. Urban classification performance is the correct classification of urban test samples only. For our application, miss rates in urban region detection are more important than false alarms. False alarms can be detected in subsequent processing while misses can not be recovered easily.

As can be seen, we achieve a clear improvement at each step for urban and average classification performances. This information appears in tabular form in Table 7.5.

To compare our hybrid features with the structural or multispectral features alone, we take the multispectral feature performance for each classifier as a *benchmark*

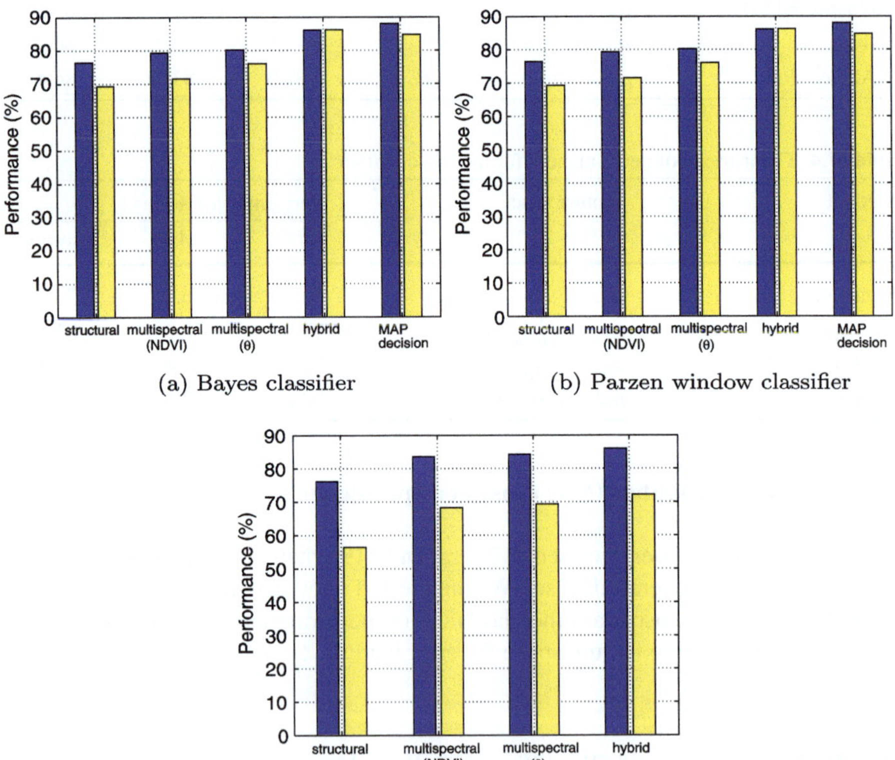


(a) Bayes classifier (b) Parzen window classifier

(c) NN classifier

**Fig. 7.8** Improvements in classification for the Bayes, Parzen window, and NN classifiers. *Dark bars* correspond to urban performances, *light bars* correspond to average performances

**Table 7.5** Comparison of detection performances for structural, multispectral, and hybrid features

| Classifier/Features | Performance (%) | | | | | |
|---|---|---|---|---|---|---|
| | Structural | | Multispectral | | Hybrid | |
| | urban | average | urban | average | urban | average |
| Bayes | 69.4 | 76.5 | 71.6 | 79.4 | 86.2 | 86.1 |
| Parzen window | 54.6 | 75.3 | 70.2 | 84.7 | 72.7 | 86.2 |
| NN | 56.4 | 76.2 | 68.3 | 83.6 | 72.3 | 86.0 |

(The structural features provide slightly poorer performance than the multispectral). The improvements in average classification rate are **8.6**%, 1.7%, and 2.5% for the Bayes, Parzen window, and NN classifiers, respectively. The respective improvements in urban region classification are **14.6**%, 12.4%, and 4.0%.

The **8.6**% improvement via hybrid features by the Bayes classifier shows the strength of our data fusion method. In the Parzen window and NN classifiers, we again have improvements. Therefore, we can conclude that our hybrid features give better performance independent of the classifier used. In addition, if the Bayes classifier is used, we get even better performance. Since we have dominant structure in urban regions, fusing them with multispectral information gives even better results, namely **14.6**% improvement.

Table 7.5 shows that multispectral features perform better than the structural features. The main reason for the relatively poor performance of the structural features is their sensitivity to the window size. In the previous chapter, using only panchromatic images, the window size was $400 \times 400$ pixels; here it is $800 \times 800$ pixels owing to the lower resolution of the multispectral images.

#### 7.4.3.2 Comparison of *NDVI* and $\theta$ Features

To compare the *NDVI* and $\theta$ features, we tabulate their performances in Table 7.6. For the average classification performance, we have around 1% improvement for all three classifiers. These improvements may not seem significant; however, there is virtually no cost to achieve them. We simply take the arctangent of the *NDVI* to obtain the $\theta$ representation. If we focus on urban region detection rates, we see significant improvements, in the range of 1.1–4.5%. These results indicate that the linearized $\theta$ representation improves classification performance compared to the *NDVI*.

#### 7.4.3.3 Comparison of Statistical and MAP Classifiers

Finally, let us consider the effect of injecting spatial information into the classification process. We achieve this by taking MAP decision after relaxation explained in Sect. 7.3. The average classification rates show **1.8**% and 0.2% improvements by using the MAP decision over the hybrid features for the Bayes and the Parzen

**Table 7.6** Comparison of detection performances for the *NDVI* and $\theta$ based features

| Classifier/Features | Performance (%) | | | |
|---|---|---|---|---|
| | *NDVI* | | $\theta$ | |
| | urban | average | urban | average |
| Bayes | 71.6 | 79.4 | 76.1 | 80.2 |
| Parzen window | 70.2 | 84.7 | 71.4 | 85.5 |
| NN | 68.3 | 83.6 | 69.4 | 84.3 |

window (pre)classifiers. These improvements are marginal when we take the computational cost of relaxation into account. For urban region detection rates, we see a drop of 1.4% for the Bayes classifier, but an **11.1%** improvement for the Parzen window classifier. Although we suffer a deterioration in the Bayes classifier, the improvement in the Parzen window classifier result is remarkable. This improvement is mainly due to the characteristics of the Parzen window classifier, and the feature space it has.

### *7.4.4 Analysis of Misclassification Results*

One of the main source of misclassification for all three classifiers is the transitions between urban and rural regions. Unfortunately, some of our images lie on these transition regions. However, for an automatic region classification system this problem is unavoidable, since the land property is unknown a priori. A sliding window approach with an intelligent feature assignment strategy could mitigate this problem. This lies beyond the scope of this book.

The second main source of misclassification is the phantom structure formations due to shadows in satellite images. These phantom structures have significant effect on structural features extracted from rural regions. Also for the multispectral feature extraction, there may be rock formations having similar characteristics as in urban regions. There may also be highways and farms (taken as rural for our problem) in rural regions, misclassified due to the definition of urban and rural regions for our purposes.

## 7.5 Summary of the Chapter

In this chapter, we improved our previous results in assessing the degree of land development in three steps. Our first step was to use the linearized *NDVI*, which we call $\theta$, as the source of measurement for multispectral information. Our second step was to combine the structural and the multispectral information by conditioning the extraction of vegetation indices on the presence of a corresponding line support region in the panchromatic image. Our last step was to enforce spatial coherence in classification through probabilistic relaxation.

Our most promising features are the hybrid features, combining the structural and multispectral information. These two information sources reflect different properties of the image they depend on. Therefore, their combination improves classification.

Compared to the *NDVI*, the normalized angle $\theta$ also improved classification. Besides being more linear, this classification improvement also indicates the usefulness of $\theta$ for general vegetation index studies.

By using parametric (Bayes) and non-parametric (Parzen window and NN) classifiers, we were able to distinguish developed areas with high confidence. Of course, our ground truth may reflect the actual development imperfectly because of human inconsistencies in assessing the ground truth labels. (A single individual provided all "ground truth" labels to minimize the inconsistencies as much as possible.) We speculate that our results could be better than given here, were the ground truth to be constructed by expert image analysts adhering to a strictly-defined standard.

Use of the spatial information by relaxation further improved the classification performances. However, this improvement comes at an additional computational cost. Therefore, relaxation should be confined to where the performance is of utmost importance.

## 7.6 Problems

**7.1** Where does probabilistic relaxation fit within the groups of land use classification?

**7.2** How is Mahalanobis distance different from the Euclidean distance?

**7.3** Compare structural, multispectral, and hybrid features in terms of performance, computation cost, and data requirements.

**7.4** What is the main misclassification reason for features in this chapter?

**7.5** Why does $\theta$ have a better performance than the *NDVI*?

**7.6 (Open ended question)** Can other hybrid features be introduced based on the information in this chapter?

## References


1. J.R.G. Townshend, C. Justice, W. Li, C. Gurney, J. McManus, Remote Sens. Environ. **35**, 243 (1991)
2. V. Caselles, E. Valor, Remote Sens. Environ. **57**, 167 (1996)
3. J. Symnazik, R.R. Griffiths, L. Gillies, in *Proceedings of the Statistical Computing Section and Section on Statistical Graphics* (2000), pp. 10–19
4. Y.J. Kaufman, L.A. Remer, IEEE Trans. Geosci. Remote Sens. **32**(3), 672 (1994)

5. T. Fung, in *Proceedings of IGARSS*, vol. 2 (1997), pp. 836–838
6. J.M.C. Pereira, IEEE Trans. Geosci. Remote Sens. **37**(1), 217 (1999)
7. P.R. Coppin, M.E. Bauer, IEEE Trans. Geosci. Remote Sens. **32**(4), 918 (1994)
8. D.F. Lozano-Garcia, R.N. Fernández, C.J. Johannsen, IEEE Trans. Geosci. Remote Sens. **29**(2), 331 (1991)
9. T. Ishiyama, S. Tanaka, K. Uchida, S. Fujikawa, Y. Yamsahita, M. Kato, Adv. Space Res. **28**(1), 183 (2001)
10. P.M. Teillet, K. Staenz, D.J. Willimas, Remote Sens. Environ. **61**, 139 (1997)
11. R.L. Kettig, D.A. Landgrebe, IEEE Trans. Geosci. Electron. **14**(1), 19 (1976)
12. A. Rosenfeld, R.A. Hummel, S.W. Zucker, IEEE Trans. Syst. Man Cybern. **6**, 420 (1973)
13. H. Yamamoto, Comput. Vis. Graph. Image Process. **10**, 256 (1979)
14. C. Ünsalan, K.L. Boyer, IEEE Trans. Geosci. Remote Sens. **42**(12), 2840 (2004)

# Chapter 8
# Graph Theoretical Measures for Land Development

In the previous chapters, we concentrated on land use classification. This chapter extends this problem and casts it as measuring organization on land. Therefore, we introduce graph theoretical measures over panchromatic images here.[1]

We extract primitives from the image, calculate measures on these primitives, and fuse these measures to determine the final region type. Our primitives are lines extracted by line support regions; see Chap. 6). We represent each straight line segment as a vertex in a graph and define a neighborhood tolerance to construct edges between these vertices. We then compute measures on these graphs to infer the type of region. These measures generally increase with respect to the degree of organization in the image. To form these measures, we first consider unweighted graphs and use the circuit rank and degree (valency) sequence. Then, we consider weighted graphs and introduce measures based on graph partitioning and the graph spectrum.

For our purposes, the level of development (or, roughly, the degree of organization) is based on the type and density of construction (buildings, streets, etc.) and its geometric regularity. Our measures are defined to infer this indirectly from the organization of lines in the image. We consider rural areas without buildings to be the least developed, proceeding through sparse residential, dense residential, commercial, industrial, to urban centers. We concede that the concept of "degree of organization" is not mathematically precise. This chapter represents a step toward quantifying this notion.

We conducted an experimental investigation to evaluate these measures. The experiment is designed to verify their approximately monotonic behavior. Because the notion of "ground truth" is speculative, at best, in this setting, we asked 90 people to sort a set of 20 test images according to their assessment of the degree of organization. We then compare our measures to the average ordering. Although each measure performed reasonably well on its own, performance improved when the measures were fused, as we describe below.

---

[1]The figures in this chapter are obtained from our previous work [1]. Here, they appear with the kind permission of IEEE.

C. Ünsalan, K.L. Boyer, *Multispectral Satellite Image Understanding*,
Advances in Computer Vision and Pattern Recognition,
DOI 10.1007/978-0-85729-667-2_8, 

There exists relatively little prior work in the direct assessment of land development from overhead views based on image structure, as opposed to pixel by pixel radiometry. Among the more notable—and philosophically most similar—is the work of Sarkar and Boyer [2] in change detection from graph spectra. Because they were looking at smaller areas such as construction sites, and because their images were not metric orthophotos (as ours are), they faced a somewhat different environment. Our graphs are simpler to construct, and our measures are potentially more discriminating.

## 8.1 Graph Construction and Consensus Ordering

In abstracting the image to a graph, we make the following observations. First, the impact of human activity, especially development, is manifested in the emergence of straight and smoothly curved contours. Second, the spatial density and regularity of these contours increases with increasing development. Here, the straight line segments defined above correspond to the vertices of the graph.

Vertices may be attributed by the length, orientation, contrast (defined in (6.14)), and centroid location of the corresponding straight line segment. Edges connect vertices corresponding to lines that are sufficiently close to one another. For the purpose of edge construction, the distance between two lines is the minimum distance between their point sets. We tested neighborhood tolerances of 3, 5, and 10 meters (pixels); experimentally we have found that 5 pixels works well. We consider the effect of different tolerance values below. An edge may be weighted according to some function of the attributes of the vertices it joins.

Each image in our data set is $800 \times 800$ pixels; we have hundreds of test images at hand. Of these, we selected 20 representatives covering the range of development from wilderness/open water to dense urban. These representative images appear in Fig. 8.1, and their extracted lines in Fig. 8.2. The geographical location and the type of each sample test image is summarized in Table 8.1.

The image supplier has geometrically corrected them prior to distribution. The correction process removes image distortions introduced by the collection geometry and resamples the imagery to a uniform ground sample distance (GSD) and a specified map projection.

We asked 90 people to sort the images from least organized (score $= 1$) to most (score $= 20$). We provided no further definition or clarification of what was meant by "organized". We defined the consensus ordering according to the median scores. For regions with discernible development activity, the responses were fairly consistent. For regions with little to no development, the human subjects were less in agreement—but they were consistent in ranking these images below those showing development. Admittedly, this hardly constitutes "ground truth" in a proper scientific sense, but it does provide a reference frame for evaluation. Substituting "regularity" for "organized" may produce a different consensus.

For the sample images, we obtain the score distributions represented by the box-and-whisker plots in Fig. 8.3. The box has lines at the lower quartile, median, and

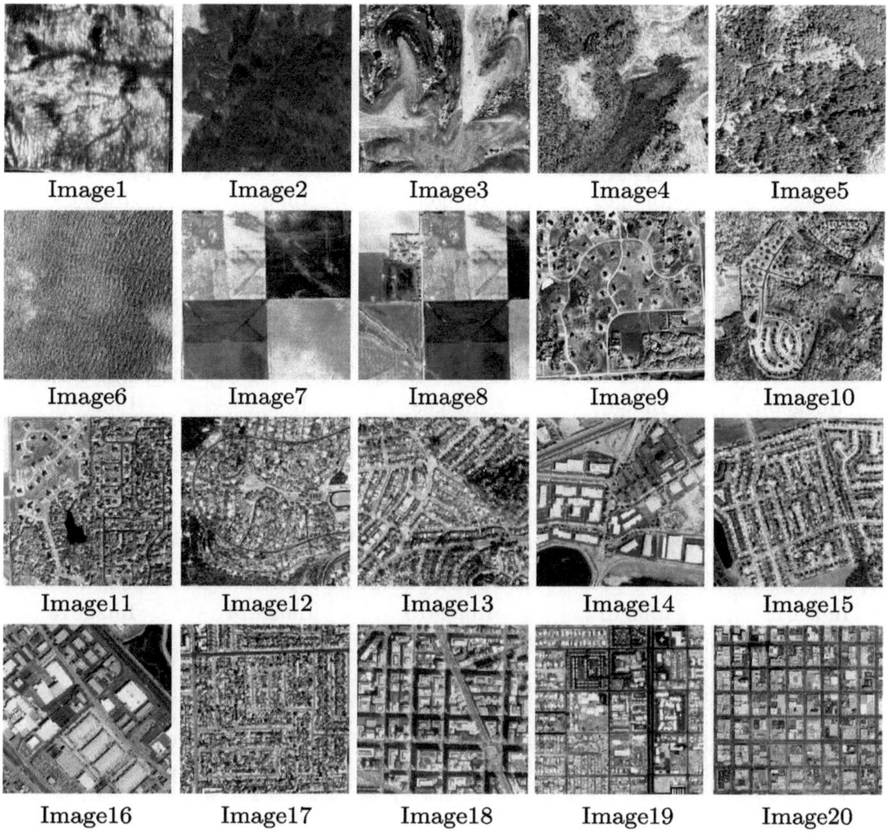


**Fig. 8.1** Sample test images arranged in increasing order by consensus ordering

upper quartile values. The whiskers are lines extending from each end of the box to show the extent of the rest of the data under a Gaussian assumption. Data values beyond the whiskers correspond to outliers. We note a clear distinction in organization scores between the sixth and seventh images.

## 8.2 Measures Based on Unweighted Graphs

For our first investigation, we consider unweighted graphs. The first measure derived from these graphs is based on the circuit rank; the second is based on the degree sequence, using a random graph formulation to model the effect of noise. These measures quantify, in some sense, the underlying structure of the graph.

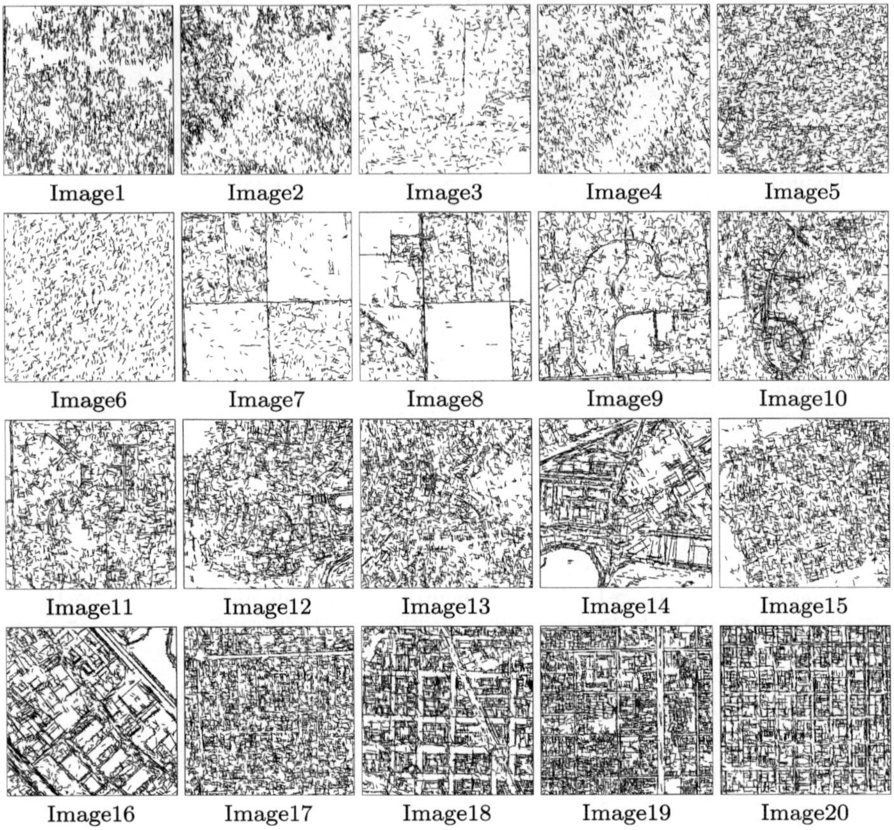


**Fig. 8.2** Lines extracted from the sample test images

### 8.2.1 Circuit Rank

Lines produced by the same house or building are likely to produce a cycle in the graph. A cycle is a closed path of alternating vertices and edges. Thus, cycles will be prevalent in graphs constructed from dense urban regions (or in the corresponding subgraphs), but rare in graphs (subgraphs) constructed from unpopulated areas. Therefore, we expect the circuit rank to increase as we move from rural regions to the city center.

Berge [3] defines the circuit rank (cyclomatic number) of graph $G$ as

$$R(G) = m - n + c, \tag{8.1}$$

where $n$ is the number of vertices, $m$ is the number of edges, and $c$ is the number of separate connected components in the graph. He proved that the circuit rank is equal to the maximum number of independent cycles. Therefore, we take the circuit rank of a graph as our first measure $m_{cr} = R(G)$.

**Table 8.1** Geographical locations and types of the sample test images

| Image # | Image name | Image type |
|---|---|---|
| 1 | Idaho | Rural |
| 2 | Minnesota | Forest |
| 3 | Kansas | Rural |
| 4 | Minnesota | Rural |
| 5 | Oregon | Forest |
| 6 | Florida | Sea |
| 7 | Oklahoma | Farmland |
| 8 | Oklahoma | Farmland |
| 9 | South Dakota | Sparse residential |
| 10 | Maryland | Sparse residential |
| 11 | Indiana | Sparse residential |
| 12 | New Mexico | Sparse residential |
| 13 | California | Dense residential |
| 14 | California | Commercial |
| 15 | Missouri | Dense residential |
| 16 | California | Industrial |
| 17 | California | Dense residential |
| 18 | Maryland | City center |
| 19 | Arizona | Dense residential |
| 20 | California | City center |

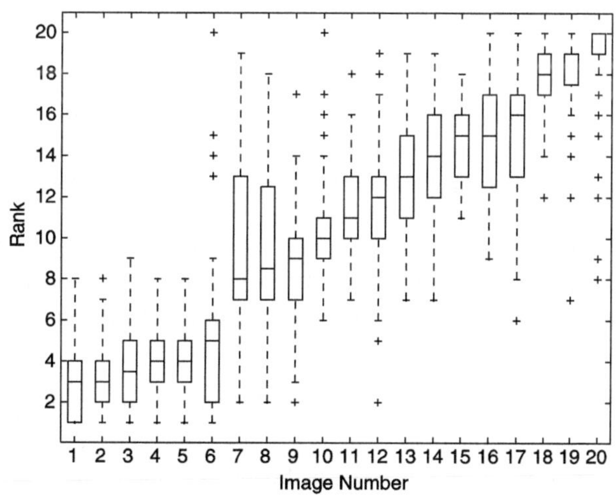


**Fig. 8.3** Consensus ordering and the score distribution in terms of box-and-whisker plots. Each box presents lines at the lower quartile, median, and upper quartile values. The whiskers are lines extending from each end of the box to show the extent of the rest of the data under a Gaussian assumption. Data points (*crosses*) beyond the whiskers are considered to be outliers

### 8.2.2 The Degree Sequence

As a region becomes more densely developed, we expect the number of connections in the graph representing it to increase. The connections may also increase in similarity, since we will have similar substructures repeated over the image. We endeavor to model this effect using the degree sequence in a random graph formulation.

Let $G$ be a graph with vertex set $V = \{v_1, v_2, \dots, v_n\}$. The number of edges incident on a given vertex $v_i$ is called the *degree* of $v_i$ and is denoted by $deg(v_i)$. The list of numbers $(deg(v_1), deg(v_2), \dots, deg(v_n))$ is called a *degree sequence* of $G$. Bollobas [4] studied random graphs to obtain the distribution of their degree sequences. Following Bollobas, we use the random graph model $\mathbf{G}(n, P(edge) = p)$. The model consists of all graphs with vertex set $V$ for which edges are chosen independently and with probability $p$. Our motivation for this measure is that as the region becomes more organized and more densely developed, the edge formation probability $p$ increases along with the number of vertices $n$.

Bollobas proved that the degree of a vertex in a random graph has a binomial distribution $b(k; n-1, p)$ with parameters $n-1$ and $p$. Based on this and the following constraints, he obtained the distribution of the degree sequence for a random graph.

**Theorem 8.1** (Bollobas) *Let $\epsilon > 0$ be fixed, $\epsilon n^{-3/2} \le p = p(n) \le 1 - \epsilon n^{-3/2}$, let $k = k(n)$ be a natural number, let $X_k$ be the number of vertices of degree $k$, and set $\lambda_k = \lambda_k(n) = nb(k; n-1, p)$. Then, the following assertions hold*:

(i) *If* $\lim_{n\to\infty} \lambda_k(n) = 0$, *then* $\lim_{n\to\infty} P(X_k = 0) = 1$.
(ii) *If* $\lim_{n\to\infty} \lambda_k(n) = \infty$, *then* $\lim_{n\to\infty} P(X_k \ge t) = 1$ *for every fixed* $t$.
(iii) *If* $0 < \limsup_{n\to\infty} \lambda_k(n) \le \liminf_{n\to\infty} \lambda_k(n) < \infty$ *then* $X_k$ *is asymptotically Poisson distributed with mean* $\lambda_k$:

$$P(X_k = r) \approx \frac{e^{-\lambda_k} \lambda_k^r}{r!} \quad \textit{for every fixed } r. \tag{8.2}$$

For the proof of this theorem, see [4].

The conditions in the theorem hold as $n$ goes to infinity. In our domain, $n$ is sufficiently large to find an approximate distribution. We will see the effect of this approximation while forming the actual (normalized) degree sequence histograms next.

For our problem $n \approx 2000$, $k \approx 20$ (typical vertex has degree 20), and $p \ll 1$. Therefore, we can use the Poisson approximation for the binomial distribution for the $\lambda_k$ (mean of the Poisson distribution for the number of vertices of degree $k$) as:

$$\lambda_k = nb(k; n-1, p) \approx \frac{n e^{-\lambda_p} \lambda_p^k}{k!}, \tag{8.3}$$

where $\lambda_p = (n-1)p$, $\lambda_p$ is the mean of the Poisson distribution of $\lambda_k$ divided by $n$.

To construct the measure, we need the edge probability $p$. To estimate $p$ from the distribution of $X_k$, we use the expectation:

$$E\left[E[X_k/\lambda_k]\right] = n\lambda_p = n(n-1)p. \tag{8.4}$$

Using the conditional expectation in (8.4), we obtain the product of $n(n-1)$ and $p$. Since each offers some indication of the organization in the graph, their product serves as the first of two measures based on the degree sequence.

Four sample degree sequence distributions are shown in Fig. 8.4. These images cover the full range of development. The normalized degree sequence histograms we obtain are very close to the Poisson distribution. In these figures, we give the normalized histogram of the degree sequence of each image and the best Poisson distribution fit. To fit the Poisson distribution, we search for the distribution mean that minimizes the squared error between the theoretical distribution and the histogram. In these figures, filled circles indicate the actual normalized histogram values and

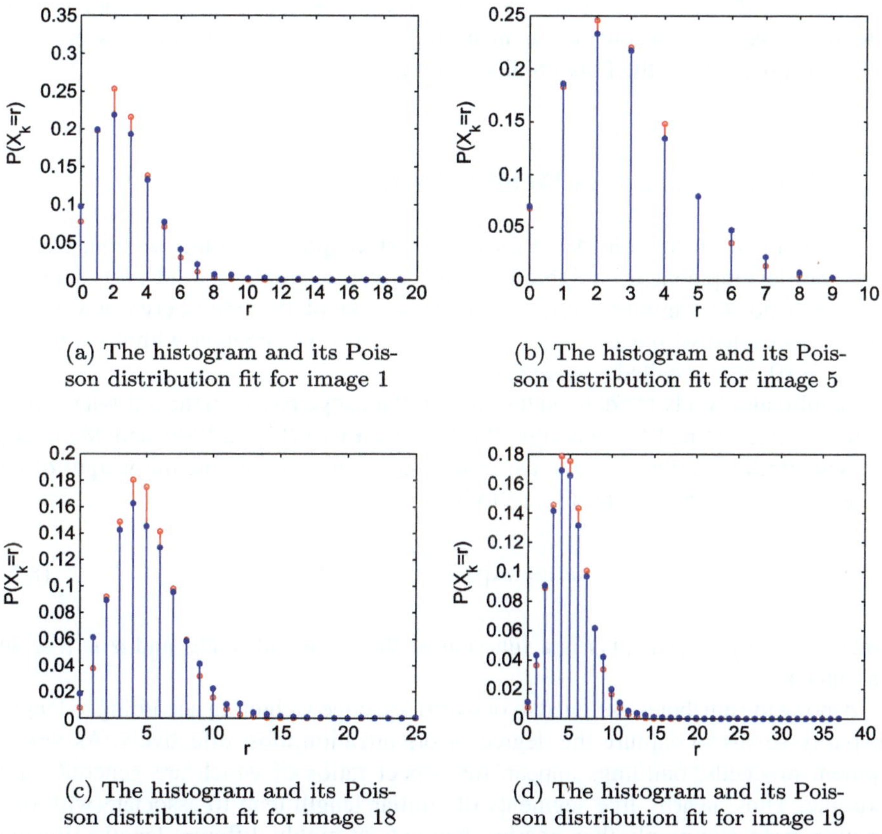


(a) The histogram and its Poisson distribution fit for image 1

(b) The histogram and its Poisson distribution fit for image 5

(c) The histogram and its Poisson distribution fit for image 18

(d) The histogram and its Poisson distribution fit for image 19

**Fig. 8.4** Sample degree sequence histograms and their best Poisson distribution fits. *Filled circles*—actual histograms, *empty circles*—Poisson distribution values the empty circles represent the corresponding theoretical Poisson distribution values.

The normalized histograms and the Poisson distributions fit fairly well. This indicates that Bollobas' theorem applies reasonably well to our case (although $n$ does not go to infinity). Therefore, our first measure derived from the degree sequence $m_{dsf}$ is the mean of the Poisson distribution as calculated from a least squares fit.

Although this measure is reasonably informative, we have observed that if we estimate this mean by sample moments, we get better agreement with the human ordering. We can estimate the mean of the Poisson distribution by dividing the square of the first moment by the second central moment. This ratio simplifies to the mean for the Poisson distribution. Therefore, the second measure derived from the degree sequence $m_{ds}$ is:

$$m_{ds} = \frac{\mu_{\text{deg}}^2}{\sigma_{\text{deg}}}. \tag{8.5}$$

This estimate of the mean is slightly superior to the other. One reason for this may be that this calculation also accounts for fit errors, using all the data more effectively. Another benefit of this measure is that it does not require a search for the least squares fit to the Poisson distribution.

## 8.3 Measures Based on Weighted Graphs

To this point we have considered only unweighted graphs. In this section, we extend the development to weighted graphs. Given a graph $G = (V, E)$, suppose there is a positive number $w_{ij}$, called the weight of the edge between vertex $i$ and $j$, associated with each edge $e_{ij} \in E$. The graph $G$, together with the function $w : E \to (0, \infty)$, is called a *weighted graph*.

Shapiro and Brady [5] defined the weight of an edge based on the attribute values of the vertices joined by that edge. Sarkar and Boyer [2] and Shi and Malik [6] defined similar weighting methods. In the same spirit, we define the weight of an edge $e_{ij}$ between two vertices $v_i, v_j$ to be:

$$w_{ij} = \exp\left(\frac{-\beta(v_i, v_j)}{\alpha}\right), \tag{8.6}$$

where $\beta(v_i, v_j) = \beta(v_j, v_i)$ is a function of the vertex attributes and $\alpha$ is a scale parameter.

Among the attributes assigned to our vertices, edge weighting according to length disparity seems to capture the degree of organization most effectively. As development proceeds, buildings appear, the aspect ratios of which are generally not extreme. Thus, nearby line segments of similar length tend to associate with one another more commonly than nearby segments of highly different lengths. Basing the edge weights on the similarity in length between the two corresponding line segments captures this observation. So, we set $\beta(v_i, v_j) = |l_i - l_j|$ where $l_i$ is the length attribute of vertex $v_i$. After extensive testing, we settled on $\alpha = 50$ as a suitable scale parameter. (Recall that we deal with metric images, so a fixed scale is suitable.) We explored the performance of different attributes, as well as the effect of the scale $\alpha$, on the final result in the following sections.

The weighted *adjacency matrix* of $G$ is the $n \times n$ matrix $\mathbf{A} = [w_{ij}]$. The eigenvalues and eigenvectors of $G$ are those of $\mathbf{A}$. If $\lambda_1, \ldots, \lambda_n$ are the eigenvalues of $G$, the *spectrum of the graph* is defined as $sp(G) = [\lambda_1, \ldots, \lambda_n]$.

The *degree matrix* $\mathbf{D}$ is a diagonal matrix with $\sum_j w_{ij}$ of vertex $i$ at position $(i, i)$. The *Laplacian matrix* of the graph $G$ is defined as $\mathbf{L} = \mathbf{D} - \mathbf{A}$. The Laplacian matrix is explained in detail in [7].

### *8.3.1 Graph Partitioning by the Laplacian Cut*

We next introduce three measures derived from graph partitioning, all based on a common idea: As development occurs, the number of edges in the graph (relative to the number of vertices) is expected to increase, as more linear structures in the image appear in mutual proximity. In rural regions, graph vertices are, to a great extent, isolated; indeed, the graph may consist largely of small disjoint subgraphs. Therefore, we anticipate a few large vertex clusters in urban areas versus many small(er) clusters in rural areas.

We obtain graph clusters from the Laplacian cut. Fiedler [8–10] introduced graph partitioning using the eigenvector corresponding to the second smallest eigenvalue of the Laplacian matrix; this eigenvector is called the *Fiedler vector* in the literature.

Before giving Fiedler's theorems on partitioning, we review other work using graph partitioning by eigenvectors from the computer vision literature. Sarkar and Boyer [2] used a graph partitioning method based on the adjacency matrix for change detection. Perona and Freeman [11] used a similar idea for image segmentation. Shi and Malik [6] introduced the normalized cuts method, based on the normalized Laplacian matrix, for image segmentation. Weiss [12] compared existing graph theory based segmentation algorithms. Gdalyahu et al. [13] worked on image segmentation with a similar methodology. In recent work, Pavan and Pelillo [14] have developed a framework for image segmentation based on a new graph-theoretic clustering formulation. They propose a novel combinatorial concept, the dominant set, that generalizes the notion of a maximal clique to weighted graphs.

For our application, the Laplacian cut is easy to compute and produces good results. Eigenclustering may be considered a sort of "dual" approach (in that it groups, but does not cut per se) [2]. In the present case, the Laplacian cut can be computed in less than 1% of the time required for eigenclustering, and is no less informative.

We next present Fiedler's theorems on graph partitioning. Fiedler defines the second smallest eigenvalue of the Laplacian matrix as the algebraic connectivity of the graph.

**Theorem 8.2** (Fiedler) *Let the second smallest eigenvalue $a(G)$ of the Laplacian graph $G$, $L(G)$, be called its algebraic connectivity. It satisfies*

$$a(G) = \min_{\mathbf{x} \in S} \mathbf{x}^T \mathbf{L}(G)\mathbf{x},$$

*where $S = \{\mathbf{x} = (x_1, \ldots, x_n)^T \in R_n, \sum_{i=1}^n x_i = 0, \sum_{i=1}^n x_i^2 = 1\}$, $R_n$ is the space of all real column vectors with n coordinates.*

The proof follows immediately from the Courant theorem [15] since the smallest eigenvalue of $L(G)$ is zero and the corresponding eigenvector is $e$. Thus $S$ consists of all unit vectors orthogonal to $e$, which means that $a(G)$, as the second smallest eigenvalue, is the minimum of $\mathbf{x}^T \mathbf{L}(G)\mathbf{x}$ on $S$.

Fiedler also proved that a weighted connected graph can be partitioned into two connected subgraphs based on algebraic connectivity:

**Theorem 8.3** (Fiedler) *Let $G$ be a finite connected graph with n vertices $1, \ldots, n$ to every edge $(i, k)$ of which a positive number $w_{ik}$ is assigned. Let $y = y_i$ be a characteristic valuation (eigenvector corresponding to the algebraic connectivity) of $G$. For any $r \geq 0$, let $M(r) = \{i \in N | y_i + r \geq 0\}$. Then the subgraph $G(r)$ induced by $G$ on $M(r)$ is connected.*

*Remark* A similar statement can be proved for $r \leq 0$ and the set $M'(r)$ of all those $i$'s for which $y_i + r \leq 0$. For the proof of this theorem, see [10].

Our graphs satisfy the above constraints and we define the Laplacian of a graph as Fiedler does. Therefore, Fiedler's theorems are directly applicable to our problem.

Although Fiedler did not give an optimal value of $r$ for partitioning, the median of the Fiedler vector entries has been extensively used in the literature for this purpose. In this chapter, we use this value to cut the graph into two parts. We apply the successive partitioning method suggested by Shi and Malik to obtain the final partitioning. In our implementation, the partitioning operation stops as the sum of the weights on the Fiedler vector exceeds a certain limit. We check for the energy of the Fiedler vector by calculating the median value of the weights on them. If the median of the Fiedler vector is smaller than 0.71, we stop partitioning.

#### 8.3.1.1 The First Measure on Graph Partitioning

The first of our measures based on graph partitioning is related to the size of the partitions (number of vertices in each). For well organized regions, we expect fewer clusters with more vertices in each. For rural regions, we expect more clusters with fewer vertices in each. As we checked the cluster count for graphs obtained from rural regions (with sparse lines), we observed that the largest cluster has approximately 20 vertices on average. Therefore, we define clusters having more than 20 vertices as large. We explore the robustness of the proposed measure on this constant in the following sections. The first measure $m_{lc1}$ is the fraction of vertices appearing in large clusters:

$$m_{lc1} = \frac{\sum_{i=1}^{K} |vc(i)|}{n}, \tag{8.7}$$

where $|vc(i)|$ is the number of vertices in cluster $i$ and $|vc(i)| > 20$ for $i = 1, 2, \dots, K$. Here $K$ is the number of large clusters, and $n$ is the total number of vertices.

#### 8.3.1.2 Second and Third Measures on Graph Partitioning

Our second and third measures combine structural (cluster) information with the unary attributes assigned to vertices. We use length and contrast attributes to form the measures $m_{lc2}$ and $m_{lc3}$:

$$m_{lc2} = \frac{\sum_{i=1}^{n} l(i)}{N_p}, \tag{8.8}$$

$$m_{lc3} = \frac{\sum_{i=1}^{n} c(i)}{N_p}, \tag{8.9}$$

where $N_p$ is the number of partitions, $l(i)$ is the length assigned to vertex $v_i$, and $c(i)$ the contrast (given in (6.14)) assigned to vertex $v_i$. These measures are motivated by the following observation: As development ensues, line segments grow longer and exhibit greater contrast, while being grouped into a smaller number of (larger) clusters.

### *8.3.2 Singular Values of the Adjacency Matrix*

Our last two measures use the singular values of the weighted adjacency matrix. The first measure considers the distribution of the singular values; the second considers the energy of the graph.

Since the adjacency matrix $\mathbf{A}$ is symmetric, its singular values are equal to the absolute values of its eigenvalues. Considering the size of our matrices, calculating singular values is more computationally stable than calculating eigenvalues.

#### 8.3.2.1 Circularity of the Singular Value Distribution

Wigner [16–18] introduced "the semicircle law", proving that the eigenvalue distribution of a random symmetric matrix approaches a semicircular shape as the dimension of the matrix goes to infinity. Arnold [19, 20] extended Wigner's work and obtained the same law for deterministic symmetric matrices.

**Theorem 8.4** (Arnold) *Let $\mathbf{A_n} = (a_{ij})$, $1 \leq i, j \leq n$, be the nth section of an infinite Hermitian matrix, $\{\lambda_k^{(n)}\}_{1 \leq k \leq n}$ its eigenvalues, and $\{u_k^{(n)}\}_{1 \leq k \leq n}$ the corresponding (orthonormalized column) eigenvectors. Let $v_n^* = (a_{n1}, a_{n2}, \dots, a_{n,n-1})$, put*

$$X_n(t) = \left[n(n-1)\right]^{-1/2} \sum_{k=1}^{(n-1)t} \left|v_n^* u_k^{n-1}\right|^2, \quad 0 \leq t \leq 1$$

*(bookkeeping function for the length of the projections of the new row $v^*$ of $\mathbf{A_n}$ onto the eigenvectors of the preceding matrix $\mathbf{A_{n-1}}$), and let finally*

$$F_n(x) = n^{-1} \left(\text{number of } \lambda_k^{(n)} \leq x\sqrt{n}, 1 \leq k \leq n\right)$$

*(empirical distribution function of the eigenvalues of $\mathbf{A_n}/\sqrt{n}$). Suppose*

(i) $\lim_{n\to\infty} a_{nn}(t)/\sqrt{n} = 0.$
(ii) $\lim_{n\to\infty} X_n(t) = Ct$, *where* $0 \leq C \leq \infty, 0 \leq t \leq 1.$

*Then $\lim_{n\to\infty} F_n = W(\cdot, C)$, where $W$ is absolutely continuous with (semicircle) density*

$$w(x, C) = \begin{cases} (2C\pi)^{-1}(4C - x^2)^{1/2} & \text{if } |x| \leq 2\sqrt{C}, \\ 0 & \text{if } |x| > 2\sqrt{C}. \end{cases} \tag{8.10}$$

For the proof of this theorem, see [20].

Shen [21] generalized the semicircle law to singular values (absolute values of the eigenvalues). Instead of the semicircle, one obtains a quarter circle when considering singular values.

The theorem holds as $n$ goes to infinity, and our typical graph size of $n \approx 2000$ is sufficient to observe similar characteristics unless the graph is sparse. If the graph is sparse, the distribution shows two peaks, around zero and one, in addition to the quarter circle shape. From these observations, our hypothesis for the first measure is that a graph obtained from rural regions (with sparse adjacency matrices) will display more sharply peaked singular value distributions than that of the denser city center. We also expect a more quarter circle like shape for singular value distributions obtained from more organized regions.

To assess "quarter circularity", we calculate a functional fit error. Let the actual distribution (histogram) for the singular values be $f(\lambda)$. We approximate the quarter circle function by a parabola $\hat{f}(\lambda)$ as:

$$\hat{f}(\lambda) = \begin{cases} a + b\lambda + c\lambda^2 & \text{if } 0 \leq \lambda \leq 3, \\ 0 & \text{if } \lambda > 3, \end{cases} \tag{8.11}$$

where polynomial parameters $a$, $b$, and $c$ are obtained by least squares. We justify approximating the quarter circle function by a parabola as follows. The distribution function in (8.10) is more precisely an ellipse; we do not need the exact shape to construct the measure; it is much easier to fit a parabola than an ellipse.

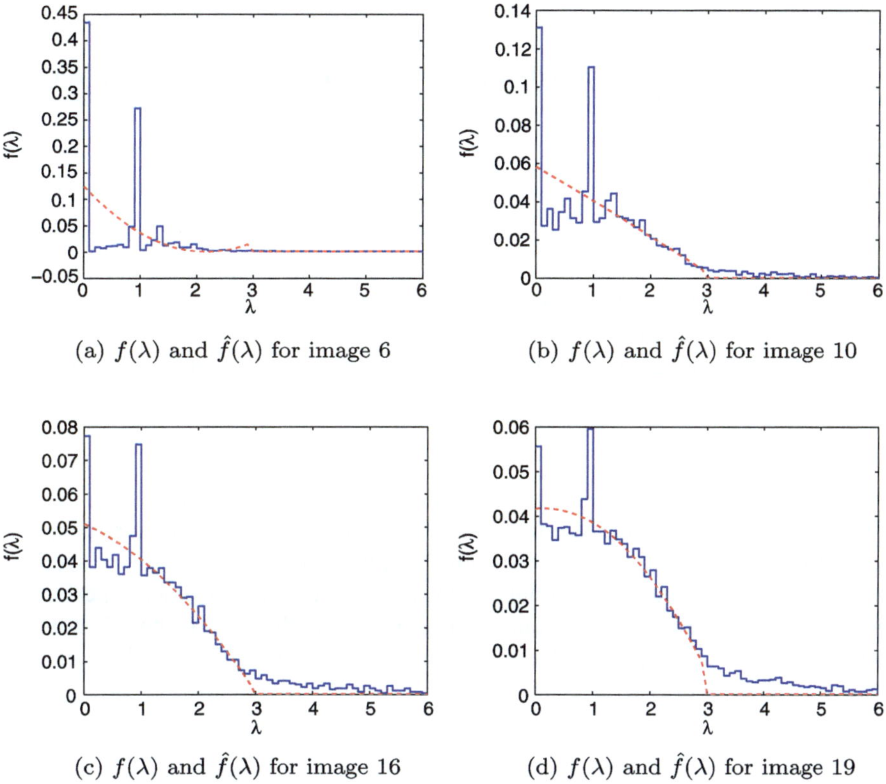


**Fig. 8.5** Normalized singular value distributions (*solid curves*) and corresponding approximations (*dashed curves*) for four sample images

We give normalized histograms $f(\lambda)$ (solid curves) and the corresponding approximations $\hat{f}(\lambda)$ (dashed curves) of singular values $\lambda$ for four sample images in Fig. 8.5. These images cover the full range of development. For the rural region (image 6) there are two dominant peaks around zero and one owing to the sparseness of the graph. The histograms of more organized regions (images 16 and 19) look more like quarter circles, as we expect. Farkas et al. [22] obtained similar distributions for small world graphs [23, 24].

The measure we propose based on quarter circularity is the negative fit error to singular value distributions:

$$m_{fe} = -\left\| f(\lambda) - \hat{f}(\lambda) \right\|, \tag{8.12}$$

where $f(\lambda)$ is the distribution of singular values and $\hat{f}(\lambda)$ its approximation. In this formulation, the least squares fit also (inversely) captures the impact of the (0, 1) peaks on the overall distribution. As the actual distribution becomes more like a quarter circle, this measure increases.

#### 8.3.2.2 The Unit Energy

The second singular value based measure is the unit energy, developed from the energy of a graph $E(G)$ as defined by Gutman and Polansky [25]:

$$E = \sum_{j=1}^{n} |\lambda_j|. \tag{8.13}$$

They introduced this definition for graphs used in organic chemistry, and it has roots in the chemical properties of molecules. Cvetkovic et al. [26] followed the same definition for general graphs.

Based on this energy definition, the unit energy measure, or energy per vertex, is:

$$m_{ueg} = \frac{E}{n}. \tag{8.14}$$

We expect the unit energy of a graph to increase as it becomes more heavily connected. This measure is derived from Sarkar and Boyer's similar measure using the sum of positive eigenvalues of the attribute graph, defined in the following section.

Although we also tested the total energy as a measure, the unit energy performed better in our experiments. The unit energy of the graph can also be taken as the sample mean of the singular values. This observation relates our second measure on singular values to the first.

## 8.4 Fusing Measures

We have defined several measures, based on different graph properties, in an attempt to capture the degree of organization and, thereby, land development. It is certainly true that "organization" and, for that matter, land development, are more than scalar quantities. For that reason, we would not use any measure alone. Therefore, we fuse them. We applied a global search over the full feature set, including all possible combinations, to construct a fused measure as consistent with human assessment as possible. (We will explain the role of the non-graph based measures $E_c$ and $\mu_c$ below.) We found that the set of the degree sequence measure $m_{ds}(i)$, Laplacian cut measure $m_{lc3}(i)$, and the distribution of singular values $m_{fe}(i)$ outperformed all other combinations.

To fuse measures having different dynamic ranges, we first normalize each measure to [0, 1] by hard limiting. A hard limiter function for each measure is constructed from the training samples by first linearly mapping the minimum value of the training set to 0.25 and the maximum value to 0.75. Using the so derived function for mapping the testing data, we then impose a hard limiter such that all values are restricted to the 0–1 range for input to the fusion operation. The final fused measure $m_f(i)$ is then the median of the normalized, limited measures.

The entropy of contrast $E_c$ is not fused directly with the graph theoretical measures. Rather, it is used in a preclassification step (below) to remove images having very little or no development (wilderness, water) from further consideration. This is certainly not optimal, and requires further investigation, but we offer this as an illustration of the different types of image assessment provided by the earlier non-graph theoretical measures versus the graph theoretical measures that are the focus of this chapter. It is a limitation of the graph theoretical measures that, while effective in grading development in regions that are, in fact, developed to some degree, they can be misled by the textures present in wave patterns and treetops under unidirectional solar illumination. The non-graph theoretical measures $E_c$ and $\mu_c$ are better at discriminating developed from undeveloped regions, but are less effective at distinguishing, for instance, suburban regions from city centers.

## 8.5 Experimental Results

We now present an experiments on these measures, comparing them with a (collective) human assessment of organization. It is fair to say that the human notion of organization is apt to be richer than what our measures capture, but the experiments show reasonable agreement, suggesting that these measures can be effective.

To compare these measures to human assessment of organization, we selected 20 images (Fig. 8.1) covering the range from (effectively) wilderness to dense urban. We note that humans are inconsistent in ranking the least developed areas—there were six such images among the 20—so a comparison against our measures would be meaningless for these six (and we have non-graph based, structural measures to eliminate these). For the remaining 14 images, 90 people demonstrated reasonable consensus—but not unanimously. For comparison, we define two types of error: *deviation* and *false alarm*. We define the deviation for each image to be $1/2$ of the absolute difference (to avoid double counts) in the ordinal position (1 through 14) as assigned by the measure versus that of the human consensus. The sum of the per-image deviations over the set provides the score for the measure.

Since we are also interested in distinguishing rural from non-rural images, we define a false alarm to be the event in which a rural image (by human classification) is assigned a ranking above the lowest non-rural image. The consequence of a false alarm in practice is nonfatal; it simply means that we will expend additional computation on a region that will ultimately prove uninteresting.

Table 8.2 shows the deviation and false alarm counts with respect to the consensus ordering. The fused measure $m_F$ presents the best performance with respect to deviation; $m_{lc3}$ provides the best performance with respect to the false alarm rate.

In Fig. 8.6, we plot each measure versus consensus ordering for all 20 images, including the first six for which the humans exhibited no real consensus. If there were perfect agreement between the consensus ranking and a measure, its curve would be monotonic. Because we are most interested in non-rural regions, it is really the monotonicity for images 7 through 20 that matters. Over this range most measures perform fairly well. Notably, $m_F$ performs better than the rest. Image 15 is consistently rated as less organized than 14 by the measures, but is scored slightly

**Table 8.2** False alarms and deviations for measures

| Measure | Origin | False alarms | Deviation |
|---|---|---|---|
| $m_{cr}$ | circuit rank | 3 | 6 |
| $m_{dsf}$ | degree sequence (l.s. fit) | 4 | 5 |
| $m_{ds}$ | degree sequence (moments) | 4 | 4 |
| $m_{lc1}$ | first partition measure | 3 | 5 |
| $m_{lc2}$ | second partition measure | 3 | 4 |
| $m_{lc3}$ | third partition measure | 2 | 4 |
| $m_{fe}$ | distribution of singular values | 4 | 3 |
| $m_{ueg}$ | unit energy of a graph | 3 | 4 |
| $m_F$ | fusion of $m_{ds}$, $m_{lc3}$, $m_{fe}$ | 4 | 2 |

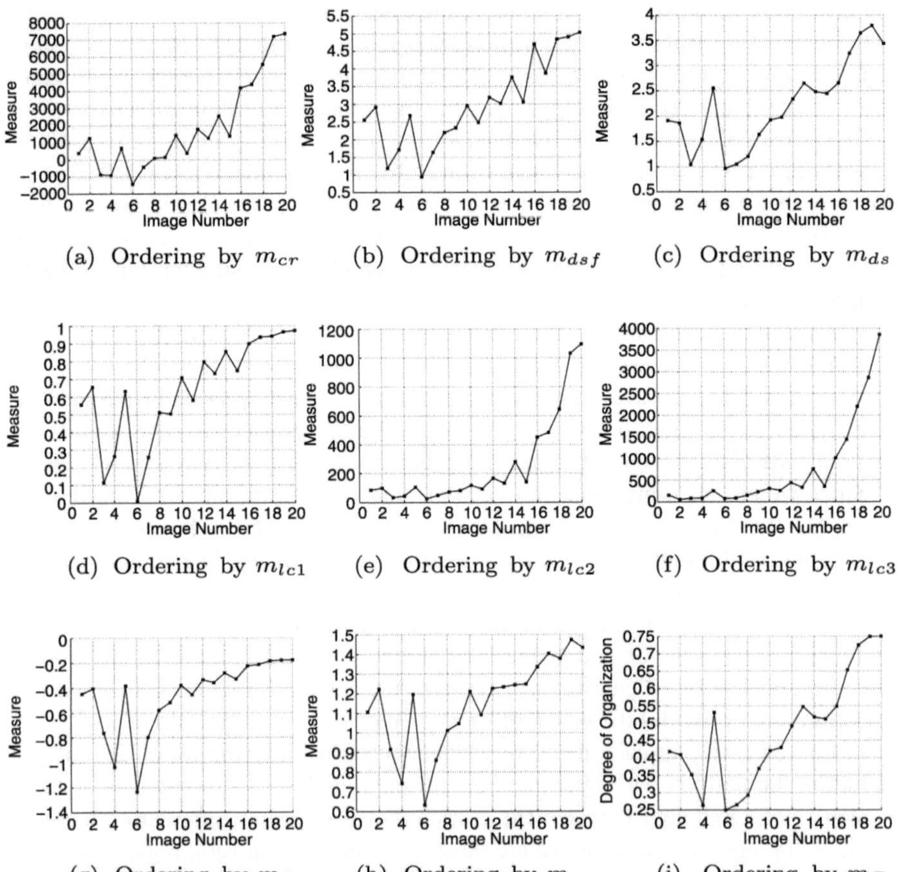


(a) Ordering by $m_{cr}$ (b) Ordering by $m_{dsf}$ (c) Ordering by $m_{ds}$

(d) Ordering by $m_{lc1}$ (e) Ordering by $m_{lc2}$ (f) Ordering by $m_{lc3}$

(g) Ordering by $m_{fe}$ (h) Ordering by $m_{ueg}$ (i) Ordering by $m_F$

**Fig. 8.6** Ordering of sample images, versus human assessment, based on all of the graph theoretical measures

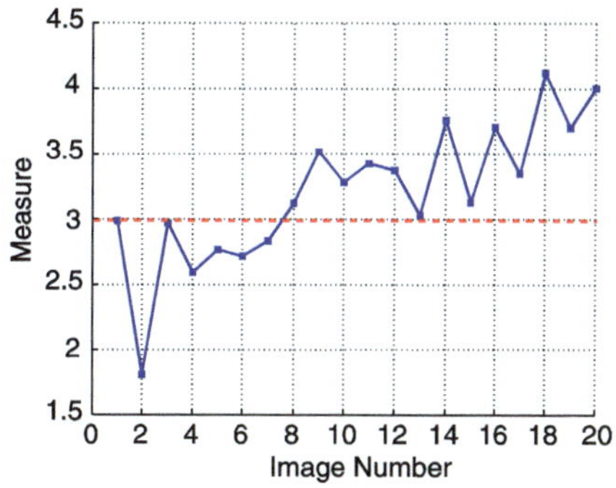


**Fig. 8.7** Eliminating the first six sample images by the entropy of contrast measure $E_c$, heavy *dashed line*—decision boundary

higher by human observers. Images 13 through 15 are very nearly equal in human assessment, so this result is not surprising.

It is instructive to consider the appearance of those "high scoring" images among the first six. These are forested regions showing phantom structure due to shadows. In these images, the sun provides unidirectional illumination imposing a fairly regular photometric structure on the image. Therefore, the edge structures in these images are nearly as regular as in residential regions. And, it is not uncommon to find significant number of trees in mature residential areas.

Figure 8.7 demonstrates a preclassifier based on the entropy of contrast $E_c$, used to eliminate the first six images. The decision boundary (shown as a heavy dashed line) indicates that a simple classifier can be used for this purpose. This is just a simple demonstration of the idea; we would be more conservative in a working system.

### *8.5.1 Sensitivity to Parameter Changes*

In this section, we consider the impact of parameter selection. For each parameter, we identify the measure most directly affected by it and tabulate the corresponding effect on the monotonicity of the measure with respect to the human ordering. This is, of course, not an exhaustive investigation of the parameter space, but provides some rationale for the values selected for these experiments and some indication of the sensitivity, vis-a-vis the human consensus.

We start with the proximity tolerance. We tabulate its effect on $m_{ds}$ in Table 8.3. A proximity tolerance value of 5 pixels seems a reasonable choice. Other tolerance values result in higher false alarms and deviations.

In another experiment, we consider the rectangularity of the cycles. Since most man-made structures comprise rectangular shapes, it is reasonable to explore such a constraint on cycle formation. We imposed a constraint that neighboring lines can participate in cycles only if they are perpendicular within a tolerance, denoted the rectangularity angle in Table 8.4.

The rectangularity constraint degrades the performance relative to the unconstrained case because we generally do not obtain precise building boundaries. The

**Table 8.3** The effect of graph neighborhood tolerance on $m_{ds}$

| Tolerance | False alarms | Deviation |
|---|---|---|
| 3 | 5 | 5 |
| **5** | **4** | **4** |
| 10 | 6 | 10 |

**Table 8.4** The effect of rectangularity constraint on $m_{cr}$

| Rectangularity angle | False alarms | Deviation |
|---|---|---|
| $\pi/32$ | 1 | 13 |
| $\pi/16$ | 1 | 10 |
| $\pi/8$ | 2 | 8 |
| $\pi/4$ | 2 | 7 |
| **no rectangularity** | **3** | **6** |

**Table 8.5** The effect of large cluster size on $m_{lc1}$

| Large cluster threshold | False alarms | Deviation |
|---|---|---|
| 5 | 3 | 5 |
| 10 | 4 | 5 |
| 15 | 3 | 6 |
| **20** | **3** | **5** |
| 25 | 3 | 7 |
| 30 | 3 | 7 |
| 40 | 3 | 9 |
| 50 | 3 | 10 |
| 100 | 2 | 10 |

image resolution is very good, but not quite good enough for smaller structures such as single-family houses to be found this way reliably, especially when overhanging trees are present.

Implicit in our definition of $m_{lc1}$ is a threshold on the number of vertices defining a *large* cluster. Our baseline value for this parameter is 20. Table 8.5 presents results for a range of values; the measure is not overly sensitive over a reasonable range, and 20 provides the best results among the values tested on our data.

In forming the weighted graph, we must select a value for $\alpha$, the normalization constant in (8.6). Table 8.6 presents the effect of this constant on $m_{lc3}$; $\alpha \geq 50$ shows good performance, and little sensitivity.

The other consideration in the construction of the weighted graph is the selection of attributes to enter the calculation in (8.6). In Table 8.7, we consider a range of attribute spaces; perhaps surprisingly, (the disparity in) the length of two line segments leads to the most effective weighting function.

**Table 8.6** The effect of $\alpha$ on $m_{lc3}$

| $\alpha$ | False alarms | Deviation |
|---|---|---|
| 10 | 4 | 16 |
| 20 | 3 | 15 |
| 30 | 3 | 8 |
| 40 | 3 | 6 |
| **50** | **2** | **4** |
| 60 | 2 | 5 |
| 70 | 2 | 4 |
| 80 | 2 | 5 |
| 90 | 2 | 5 |
| 100 | 2 | 5 |

**Table 8.7** The effect of different attributes on $m_{lc3}$

| Attribute(s) | False alarms | Deviation |
|---|---|---|
| **Length** | **2** | **4** |
| Contrast | 3 | 7 |
| Slope | 2 | 7 |
| Distance | 0 | 9 |
| Length and Contrast | 3 | 10 |
| Contrast, Slope, and Distance | 1 | 15 |
| Length, Contrast, Slope, and Distance | 1 | 18 |

### 8.5.2 Comparison with Sarkar and Boyer's Measures

The measures of Sarkar and Boyer, used for change detection, are perhaps the closest to ours to be found in the literature. In this section, we compare their measures with those proposed in this chapter. Sarkar and Boyer define their graphs on constant curvature edge segments in the image (unlike our straight lines). They obtain the weights in the graph by Gestalt-inspired relationships among vertices.

Instead of partitioning the graph as we did, they define their measures on eigenclusters of the adjacency matrix. An eigencluster is a collection of nodes corresponding to the nonzero components of a positive eigenvector ($\mathbf{x}_i$ such that all the components have the same sign).

Their first measure is $I^c_{\text{tot}}$:

$$I^c_{\text{tot}} = N \frac{\sum_{i=1}^{Nc} (\sum_{j=1}^{ni} l_{ij})}{N_c}, \tag{8.15}$$

where $l_{ij}$ is the length of the $j$th edge segment in the $i$th eigencluster, $N$ is the total number of edge segments, and $N_c$ is the total number of eigenclusters. This measure captures the degree to which edge points appear in eigenclusters.

**Table 8.8** False alarms and deviations for Sarkar and Boyer measures

| Measure | False alarms | Deviation |
|---|---|---|
| $I^c_{\text{tot}}$ | 4 | 6 |
| $-N_c$ | 1 | 14 |
| $\lambda_+$ | 4 | 10 |

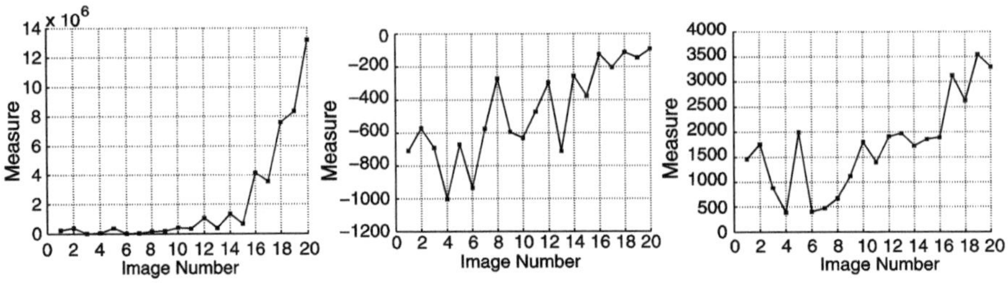


(a) Ordering by $I^c_{tot}$ (b) Ordering by $-N_c$ (c) Ordering by $\lambda_+$

**Fig. 8.8** Ordering of sample images based on three Sarkar and Boyer measures

Their second measure is the total number of eigenclusters $N_c$. For our purposes, the fewer eigenclusters a graph has, the more organized it is (assuming that the eigenclusters are also getting larger). Therefore, we use the negative of this measure for comparison.

Their last measure is the sum of the positive eigenvalues $\lambda_+$. This corresponds to a half of the energy $E$, as given in (8.13) [27].

To compare these measures with ours, we implemented the eigenclustering algorithm on our weighted graphs. Because their nodes correspond to constant-curvature arcs of edge points and ours to straight line segments, and because they account for scale and we need not, a one-to-one comparison between the two sets of measures is not possible. We are, in effect, using their measures in a manner for which they were not designed. It is nevertheless interesting to see how the Sarkar–Boyer measures behave in our situation. We give the false alarm and deviation counts for these measures in Table 8.8.

Not surprisingly, these measures are not as effective in this setting. In Fig. 8.8, we plot Sarkar and Boyer's measures versus consensus ordering for all 20 images.

## 8.6 Summary of the Chapter

In this chapter, we introduced and tested several graph theoretical measures to assess land development in overhead imagery. Our objective was to develop measures that increase monotonically with respect to the organization (development) in images, and we can report significant progress on this front. Although each measure performed fairly well on its own, the measure obtained by fusing them performed the best of all with respect to deviation from human assessment of relative organization.

Clearly, this chapter represents only a preliminary exploration of the possible leverage to be realized using graph-theoretical measures in a domain such as this. More sophisticated graph construction and attribution, as well as additional measures, may provide improved performance. Especially interesting, and important, would be subclass discrimination into different types of residential areas, or into industrial versus commercial developments. Graph embedding techniques in which the graphs are represented as points in a linear vector space also merit investigation in this domain.

## 8.7 Problems

**8.1** What is the main advantage of graph representation for land use classification?

**8.2** Why is Theorem 8.1 applicable to the problem in this chapter?

**8.3** Discuss the properties of the Poisson distribution.

**8.4** What is the relation between singular values and the eigenvalues of a matrix?

**8.5** What is the main advantage of SVD on the PCA?

**8.6** Sort the graph measures introduced in this chapter based on their computation load.

**8.7** (**Open ended question**) Can more graph-theoretical measures be introduced using the methods in this chapter?

## References

1. C. Ünsalan, K.L. Boyer, IEEE Trans. Pattern Anal. Mach. Intell. **47**(4), 575 (2005)
2. S. Sarkar, K.L. Boyer, Comput. Vis. Image Underst. **71**, 110 (1998)
3. C. Berge, *The Theory of Graphs* (Dover, New York, 2001)
4. B. Bollobas, *Random Graphs*, 2nd edn. (Cambridge University Press, Cambridge, 2001)
5. L.S. Shapiro, J.M. Brady, Image Vis. Comput. **10**, 283 (1992)
6. J. Shi, J. Malik, IEEE Trans. Pattern Anal. Mach. Intell. **22**(8), 888 (2000)
7. B. Mohar, in *Proceedings of the Sixth Quadrennial Conference on the Theory and Applications of Graphs*, vol. 2, 1998, pp. 871–898
8. M. Fiedler, Czecholoslovak Math. J. **98**, 298 (1973)
9. M. Fiedler, Czecholoslovak Math. J. **100**, 619 (1975)
10. M. Fiedler, in *Combinatorics and Graph Theory*, vol. 25 (1989), pp. 57–70
11. P. Perona, W. Freeman, in *Proceedings of the ECCV98* (1998), pp. 655–670
12. Y. Weiss, in *Proceedings of the ICCV99* (1999), pp. 975–982
13. Y. Gdalyahu, D. Weinshall, M. Werman, IEEE Trans. Pattern Anal. Mach. Intell. **23**(10), 1053 (2001)
14. M. Pavan, M. Pelillo, in *Proceedings of CVPR03* (2003), pp. 145–152

15. R.A. Horn, C.R. Johnson, *Matrix Analysis* (Cambridge University Press, Cambridge, 1985)
16. E.P. Wigner, Ann. Math. **62**, 548 (1955)
17. E.P. Wigner, Ann. Math. **65**, 203 (1957)
18. E.P. Wigner, Ann. Math. **67**, 325 (1958)
19. L. Arnold, J. Math. Anal. Appl. **20**, 262 (1967)
20. L. Arnold, Linear Algebra Appl. **13**, 185 (1976)
21. J. Shen, Linear Algebra Appl. **326**, 1 (2001)
22. I.J. Farkas, I. Derenyi, A.L. Barabasi, T. Vicsek, Phys. Rev. E **64**, 1 (2001)
23. D.J. Watts, S.H. Strogatz, Nature **393**, 440 (1998)
24. D.J. Watts, *Small Worlds, The Dynamics of Networks Between Order and Randomness* (Princeton University Press, New Jersey, 1999)
25. I. Gutman, O.E. Polansky, *Mathematical Concepts in Organic Chemistry* (Springer, Berlin, 1986)
26. D. Cvetkovic, M. Doob, I. Gutman, A. Torgasev, *Recent Results in the Theory of Graph Spectra* (Elsevier, New York, 1988)
27. G. Strang, *Linear Algebra and Its Applications*, 3rd edn. (Harcourt College Publishers, Florida, USA, 1986)

# Part IV
# Extracting Residential Regions

# Chapter 9
# Feature Based Grouping to Detect Suburbia

In this and the following chapter, we focus on detecting suburban regions among others. Although it is part of land use classification problem, we had to introduce specific methods to detect these regions in a robust manner. The direct three class classification (urban, rural, residential) approach was less successful in this case, largely because suburban regions bridge the other two in our feature space much as they do on the ground. Therefore, in an attempt to extract suburban regions, we introduced an enhancement based on the principles of perceptual organization.

Perceptual organization is that process, or a set of processes, by which a vision system (natural or artificial) organizes detected features in images based on various Gestaltic clues [1]. Perceptual organization is therefore the ability to impose structural regularity on sensory data, grouping sensory primitives having a common underlying cause. We introduced a spatial coherence constraint and performed grouping in the feature space. Via this novel perceptual grouping approach, the results improved significantly. Hence, besides the structural approach to land classification, our new spatial coherence method based on perceptual organization principles also offers very promising results by combining the feature and image spaces.

## 9.1 Feature Based Grouping

Levels of development in neighboring areas are not statistically independent. Cities tend to grow from a core outwards, apart from the intervention of natural barriers such as water and mountains, and land development expands more or less coherently. These observations indicate the importance of spatial information in classification.

To recover suburban regions, we developed a grouping and classification method motivated by principles from perceptual organization. To define the suburban, or residential class, we begin by considering the distribution of training data depicted in Fig. 9.1. This space proved to be the most effective for Bayes classification (discussed next). Clearly, the residential areas, in which we are most interested, bridge the urban and rural areas in feature space, much as they do on the ground. This

C. Ünsalan, K.L. Boyer, *Multispectral Satellite Image Understanding*, Advances in Computer Vision and Pattern Recognition,
DOI 10.1007/978-0-85729-667-2_9, 

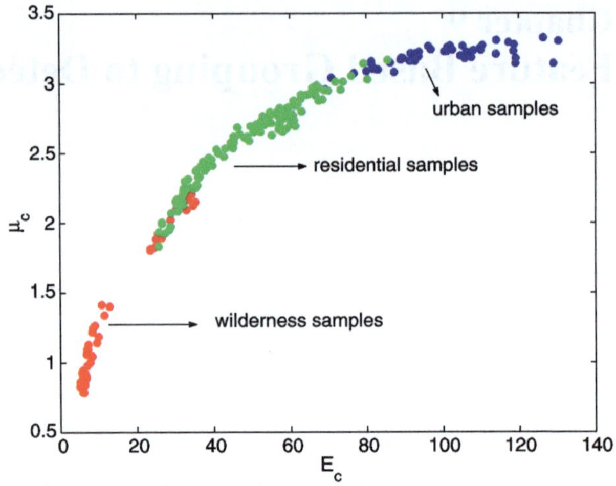


**Fig. 9.1** Sample distribution in $\mu_c$–$E_c$ space

makes the direct detection of such areas difficult, but we have been able to make significant inroads on the problem that at least suffice for our "triage".

Our attempt to build a three way Bayes classifier did not give promising results. We suspect that the feature distributions of the urban and rural classes are poorly approximated as single Gaussians. While it is theoretically possible to model these distributions as collections of multiple Gaussians (mixture models), one still needs to know how many Gaussians (subclasses) to fit, and the solution to that problem is far from obvious in the present case. Therefore, we developed an alternative approach to detect suburban regions by framing the problem as one of perceptual grouping.

Perceptual grouping is the process by which vision systems (natural or machine) impose organization on sensory data. The objective is to identify image regions (or features) arising from a common underlying cause. This exploits the fact that the world is not visually chaotic; it has structure and organization. Rather than making hard window by window decisions, we group windows together having compatible (similar) feature values into contiguous regions, and classify them together. Thus, we discover *natural groupings* of areas having similar developmental characteristics, as measured in the $(\bar{\mu_c}, E_c)$ feature space. The roots of this approach go back to Wuescher and Boyer's [2] robust contour decomposition method. Srikantiah et al. [3] implemented a 2D version of this idea to segment the surfaces of 3D free form objects. We apply this idea to more general feature spaces here.

In grouping, we first construct a 2D voting space for the feature vectors over the image. Each window votes into all bins satisfying a *compatibility* constraint. This compatibility constraint takes the form of a 2D footprint in the feature space; any window having feature values in the range covered by this footprint are considered to be compatible with the current (voting) window in that they have similar features. We then interrogate the voting result to extract contiguous regions having compatible features, from largest to smallest. The extracted collections of windows are called *segments*.

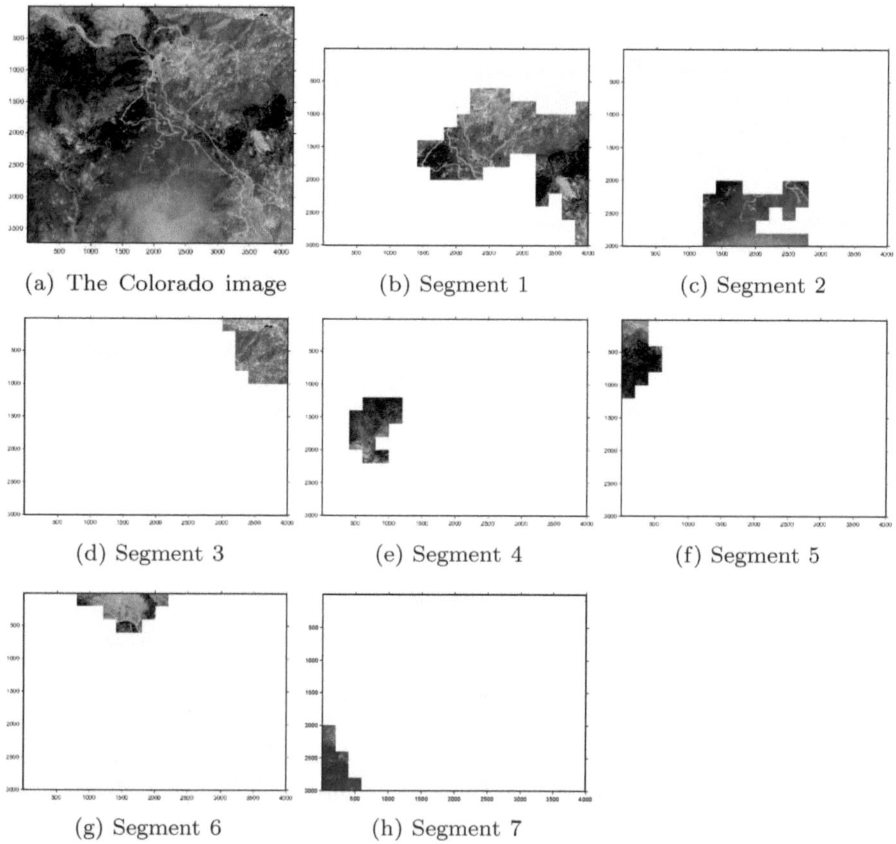

(a) The Colorado image (b) Segment 1 (c) Segment 2
(d) Segment 3 (e) Segment 4 (f) Segment 5
(g) Segment 6 (h) Segment 7

**Fig. 9.2** The Colorado image and segments obtained from feature based grouping

In these experiments, the voting spaces are of 16 bins each, uniform over the observed range of the corresponding feature value over the training data. The compatibility constraint is three bins, roughly 20% of the range. Therefore, of the $16 \times 16 = 256$ cells in the 2D voting space, each window votes for 9 cells in a $3 \times 3$ block: that corresponding to its own (quantized) feature vector plus the 8 adjoining cells. At the "corners" of the voting space, the compatibility footprint is simply truncated.

To illustrate the results of the spatial grouping algorithm, we give the Colorado, Sydney, New Mexico, and Ecuador images and the coherent segments obtained from them in Figs. 9.2, 9.3, 9.4, and 9.5.

The feature based grouping method works reasonably well in extracting homogeneously developed regions in these images. However, there are some design (and presentation) limitations. Homogeneity is computed only over square windows and is defined in the feature space. Therefore, some groupings may seem inappropriate owing to this spatial quantization. Moreover, the images are printed at different

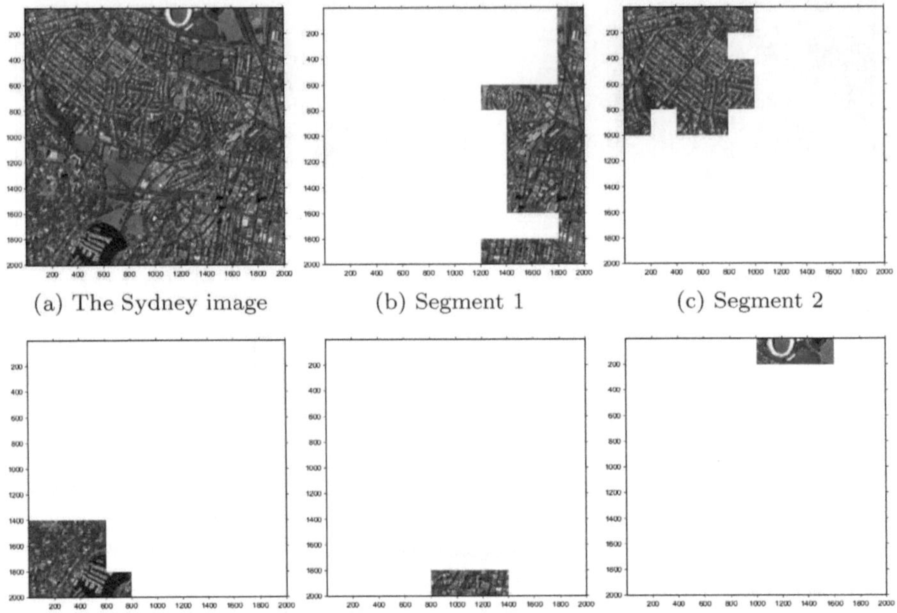

(a) The Sydney image (b) Segment 1 (c) Segment 2

(d) Segment 3 (e) Segment 4 (f) Segment 5

**Fig. 9.3** The Sydney image and segments obtained from feature based grouping

scales (due to their differing sizes). This can be misleading when evaluating the grouping visually.

To assign the final class membership to each segment, we begin by estimating the mean vector $\bar{\mathbf{x}}$ and covariance matrix $\mathbf{K}_\mathbf{x}$ of the suburban class in feature space (green points in Fig. 9.1) assuming a bivariate Gaussian distribution. We compute the Mahalanobis distance between each window to be classified and this distribution:

$$D_M = (\mathbf{x} - \bar{\mathbf{x}})\mathbf{K}_\mathbf{x}^{-1}(\mathbf{x} - \bar{\mathbf{x}})^T. \tag{9.1}$$

The degree of membership of a given window (or feature vector $\mathbf{x}$) is then computed as:

$$\mathcal{M} = e^{-(\frac{D_M}{20})}, \tag{9.2}$$

where the value 20 was selected experimentally for convenience.

We then classify a segment using $\mathcal{M}$. However, a segment's (net) feature vector is that corner of its range corresponding to each feature at its maximum. Although one might expect the average or median value of the feature range to be more representative (as we did), the "upper right" corner value performs better. The reasons are as yet unclear. All windows in the segment then receive the same degree of membership and classification.

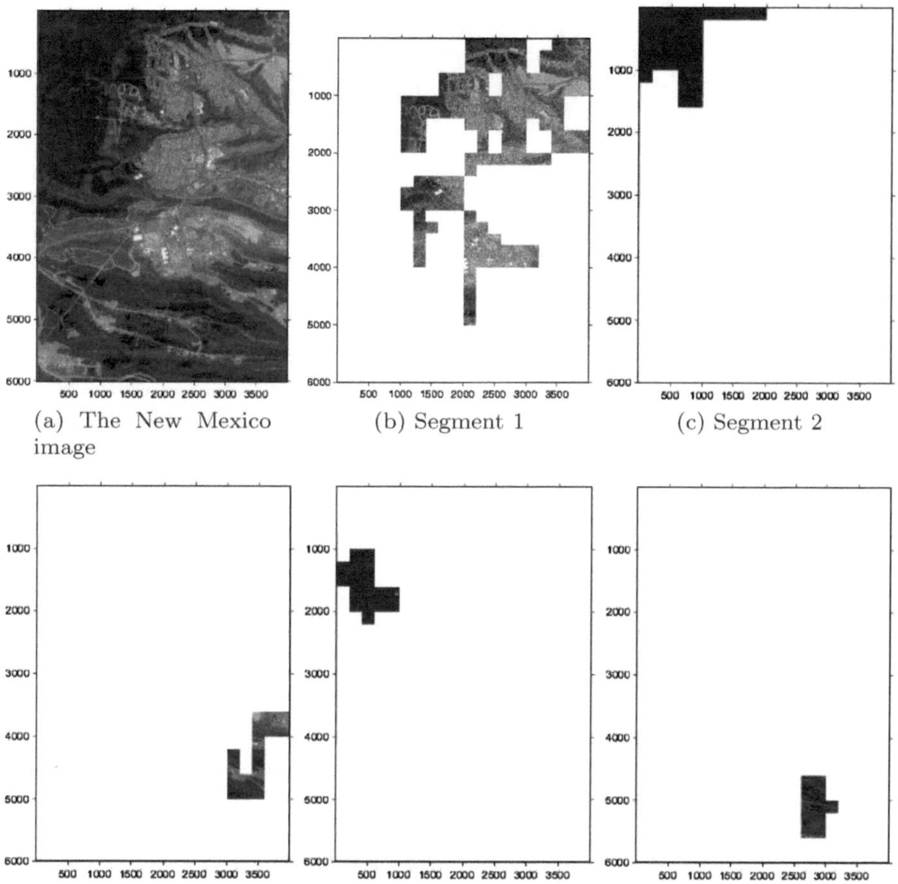


(a) The New Mexico image (b) Segment 1 (c) Segment 2

(d) Segment 3 (e) Segment 4 (f) Segment 5

**Fig. 9.4** The New Mexico image and segments obtained from feature based grouping

## 9.2 Suburban Area Detection Results

We tested the feature based grouping approach over the full data set to detect suburban regions using the membership parameter $\mathcal{M}$. For comparison, we first calculated the class membership for each sample without grouping. We used a Neyman–Pearson decision rule (specifying the false alarm rate, then obtaining the corresponding best detection rate) [4], yielding the receiver operating characteristic (ROC) labeled “Before grouping” in Fig. 9.6. This ROC was obtained by testing all of our samples. Although formulated as a two-class problem (suburbs, or not), this is really a three-class issue with fairly unequal priors (suburbs are relatively rare in this dataset), and so the ROC—while not spectacular—is far better than chance. It also makes no use (yet) of spatial coherence.

Next, we applied feature based grouping. Again, using a Neyman–Pearson decision rule, we obtained the “After grouping” ROC curve in Fig. 9.6 which reveals

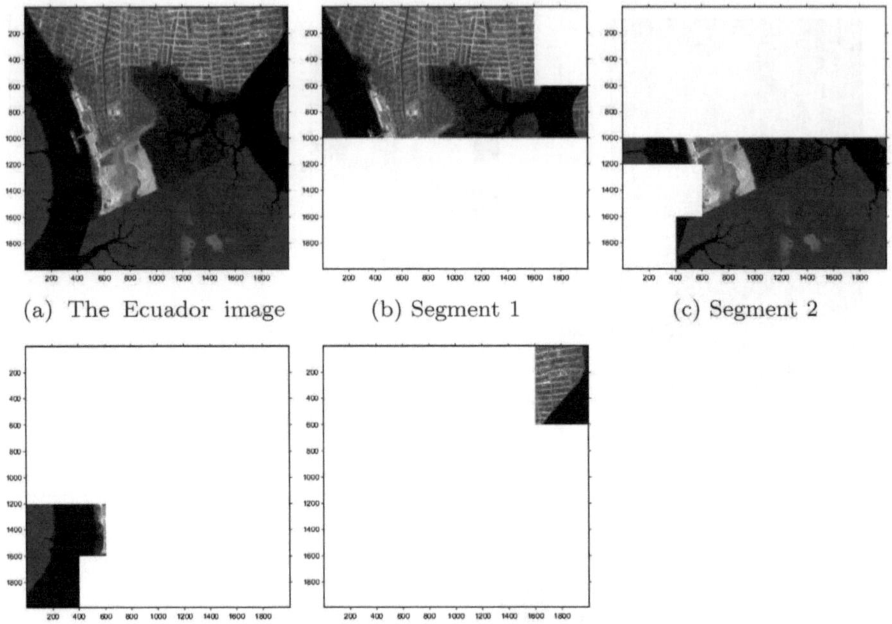

(a) The Ecuador image (b) Segment 1 (c) Segment 2

(d) Segment 3 (e) Segment 4

**Fig. 9.5** The Ecuador image and segments obtained from feature based grouping

**Fig. 9.6** ROC curves for the detection of suburban regions, with and without feature based grouping

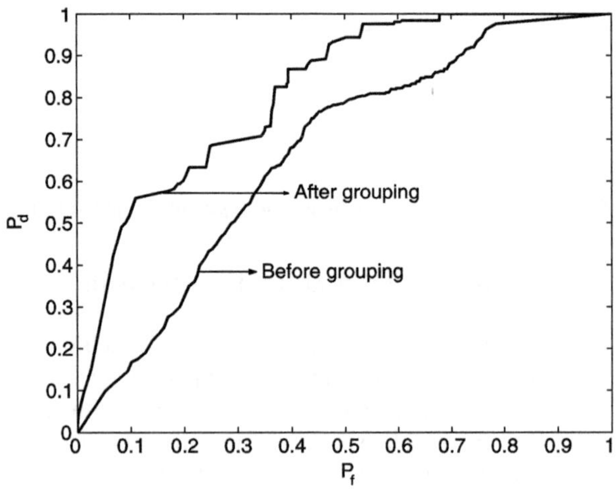


substantially better detection rates for all false alarm rates specified. This comparison shows the improvement available from feature based grouping.

The area under a ROC curve gives a single number for quality and improvement, and for the perfect case it would be one. If we compare areas under these two ROC curves, we obtain 0.6688 from the before grouping ROC curve and 0.8165 from the after grouping ROC curve. This indicates an improvement of approximately 22% by applying feature based grouping. However, it remains difficult achieving high detection rates without significant false alarms.

## 9.3 Summary of the Chapter

The detection of suburban residential regions is highly challenging. Nevertheless, a novel feature based grouping method produced highly encouraging results. This suggests that methods based on principles from perceptual organization may ultimately have significant impact on these types of problems. Although our test set is large and diverse in location and development pattern, we can distinguish developed areas with high confidence. Our results indicate that image structure, as captured by the spatial organization of its straight lines, does provide an effective indicator of land development activity.

## 9.4 Problems

**9.1** How can feature based grouping method be labeled based on the grouping in the land use classification chapter?

**9.2** What is the main hypothesis in feature based grouping?

**9.3** What do ROC curves represent?

**9.4** What are the best and worst ROC curve shapes?

**9.5** Based on the previous two questions, what is the quality of ROC curves in this chapter?

**9.6** (**Open ended question**) Can the feature based grouping method be improved?

## References

1. S. Sarkar, K.L. Boyer, *Computing Perceptual Organization in Computer Vision* (World Scientific, Singapore, 1994)
2. D.M. Wuescher, K.L. Boyer, IEEE Trans. Pattern Anal. Mach. Intell. **13**(1), 41 (1991)
3. R. Srikantiah, P.J. Flynn, K.L. Boyer, Comput. Vis. Image Underst. (2002)
4. H.V. Poor, *An Introduction to Signal Detection and Estimation*, 2nd edn. (Springer, New York, 1994)

# Chapter 10
# Detecting Residential Regions by Graph-Theoretical Measures

We next consider a more specific land class type, namely "detecting residential regions", in this chapter of the book. Unfortunately, spatial coherence based method introduced in the previous chapter is not sufficient for such fine classification. Our graph-theoretical measures, explained in Chap. 8 in full detail, can be used directly to solve this hard problem.

We focused on sparse residential areas here. These areas, often including large, overhanging trees, exhibit forest-like characteristics and present a significant challenge. Yet, they are also of particular importance because it is often in these areas that growth most rapidly occurs. To detect these regions, we implemented and tested two strategies. In the first, we developed a Neyman–Pearson decision system based on the Mahalanobis distance to the center of the residential region's distribution in our graph theoretical feature space. The second approach is to define a three-class problem (rural, residential, urban) with a Bayes classifier. We obtained very good results in both cases. It is particularly noteworthy that the residential regions show a low miss rate.

We tested 295 images taken from different parts of the US, representing a wide range of regions, climates, and terrain. We will first consider the problem of detecting residential regions only (i.e., residential vs. nonresidential classification). Then we will recast the problem as a three-way classification into rural, residential, and urban. To form (our best guess at) the ground truth, we took the consensus ordering (Fig. 8.3) as a benchmark, and labeled the test images accordingly. Unfortunately, we do not have a data set professionally labeled by experts, which would be the best way to evaluate our results. However, this labeling is true and unbiased, to the best of our knowledge and ability.

Before giving the one- and three-class classification results, let us look at some 2D histograms of the measures, given in Fig. 10.1. We label rural, residential, and urban distributions with blue, black, and red, respectively. For each figure, the $y$ axis corresponds to the first measure and the $x$ axis corresponds to the second measure, as given in the caption.

The residential region distributions lie between those of rural and urban regions, as is the case for these regions in real space. Residential regions can be separated

C. Ünsalan, K.L. Boyer, *Multispectral Satellite Image Understanding*, Advances in Computer Vision and Pattern Recognition,
DOI 10.1007/978-0-85729-667-2_10, © Springer-Verlag London Limited 2011

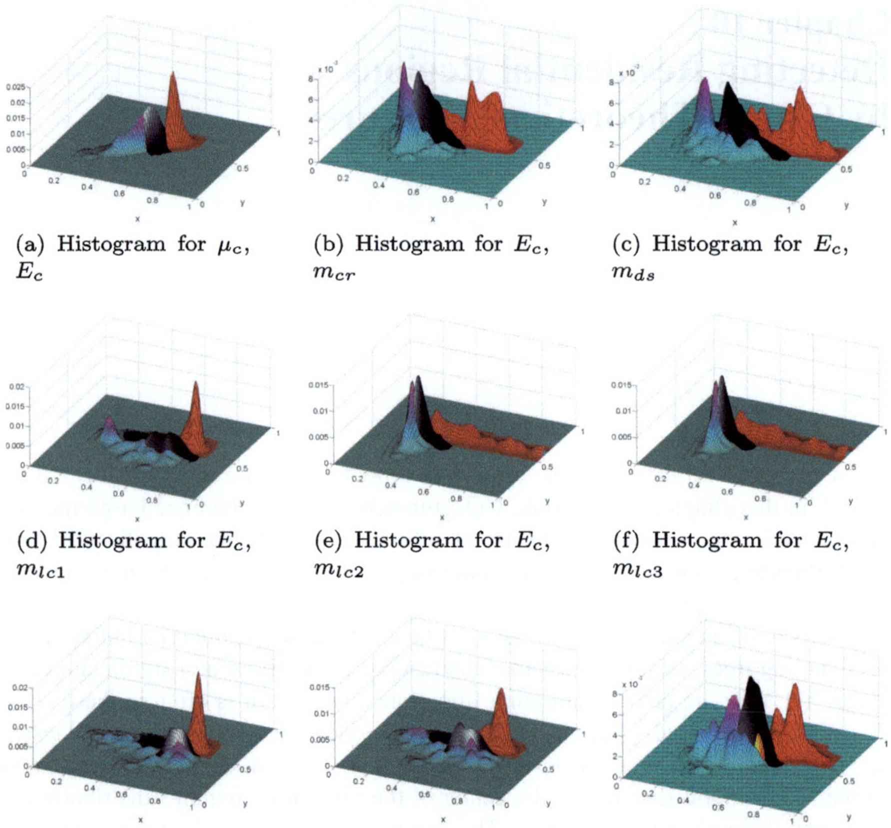

(a) Histogram for $\mu_c$, $E_c$

(b) Histogram for $E_c$, $m_{cr}$

(c) Histogram for $E_c$, $m_{ds}$

(d) Histogram for $E_c$, $m_{lc1}$

(e) Histogram for $E_c$, $m_{lc2}$

(f) Histogram for $E_c$, $m_{lc3}$

(g) Histogram for $E_c$, $m_{fe}$

(h) Histogram for $E_c$, $m_{ueg}$

(i) Histogram for $E_c$, $m_F$

**Fig. 10.1** Histograms in 2D spaces, *blue*—rural region, *black*—residential region, *red*—urban region

easily in spaces including the graph-theoretical measures. In the purely structural feature space (Fig. 10.1(a)), residential regions overlap significantly with rural and urban regions. We will see the effect of this overlap below. In most of the feature spaces, the urban regions exhibit subclusters. We do not yet know, but it will be interesting to discover, whether these subclusters correspond to meaningful subclasses of urban development.

## 10.1 One-Class Problem

Here, we formulate the problem using only residential region training data. We begin by estimating the mean vector $\bar{\mathbf{x}}$ and covariance matrix $K_{\mathbf{x}}$ of the residential region class in feature space assuming a bivariate Gaussian distribution. We then compute

**Table 10.1** Area under ROC curves

| Feature space | Area |
|---|---|
| $E_c, \mu_c$ | 0.7476 |
| $E_c, m_{cr}$ | 0.7937 |
| $E_c, m_{ds}$ | 0.8318 |
| $E_c, m_{lc1}$ | 0.8257 |
| $E_c, m_{lc2}$ | 0.8659 |
| $E_c, m_{lc3}$ | 0.8677 |
| $E_c, m_{fe}$ | 0.8428 |
| $E_c, m_{ueg}$ | 0.8061 |
| $\mathbf{E_c, m_F}$ | **0.9347** |

the Mahalanobis distance between each image to be classified and this distribution:

$$D_M = (\mathbf{x} - \bar{\mathbf{x}})\mathbf{K}_{\bar{\mathbf{x}}}^{-1}(\mathbf{x} - \bar{\mathbf{x}})^T. \tag{10.1}$$

The relative probability that a feature vector $\mathbf{x}$ belongs to a residential region is then computed as:

$$\mathcal{M} = e^{-(\frac{D_M}{10^{-3}})}, \tag{10.2}$$

where the normalization factor $10^{-3}$ is comparable with the Mahalanobis distances we observe.

We use a Neyman–Pearson decision rule. We trained the classifier using just four representative residential samples, and tested 91 residential and 190 non-residential samples. Figure 10.2 presents the receiver operating characteristic (ROC) curves for the various two dimensional feature spaces. We obtained the ROC curves by shifting the decision boundaries in the classifier. In these spaces, one dimension corresponds to the entropy of contrast $E_c$, given in (3.13), which works well in separating rural and non-rural regions; the second dimension corresponds to one of the graph based measures, which are effective at distinguishing urban from residential regions. To compare our results with the non-graph based structural feature space $(\mu_c, E_c)$ alone, we also include its ROC curve as the dashed plot in each subfigure.

Spaces of more than two dimensions performed no better; we attained our best results in the two-dimensional $(E_c, m_f)$ feature space. Table 10.1 shows the area under the ROC curve for each 2D feature space. The area under a ROC curve provides a general indication of the overall performance of that feature space. The area under a perfect ROC curve is, of course, unity. The ROC curves obtained in any space including one of the graph-theoretical measures are superior to the best available using only structural features, which is $\mu_c$, $E_c$.

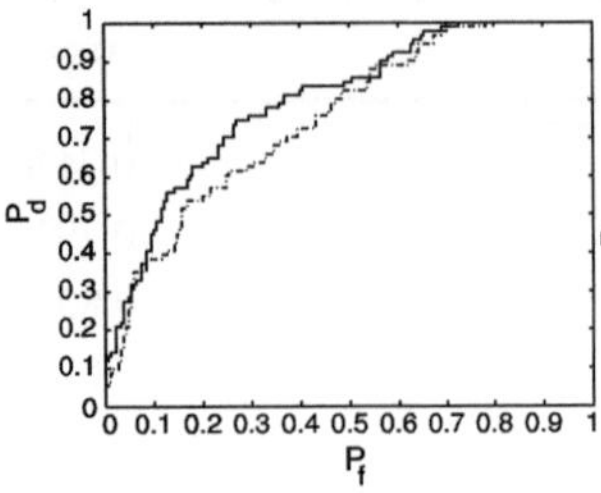


(a) ROC curve for $E_c$, $m_{cr}$

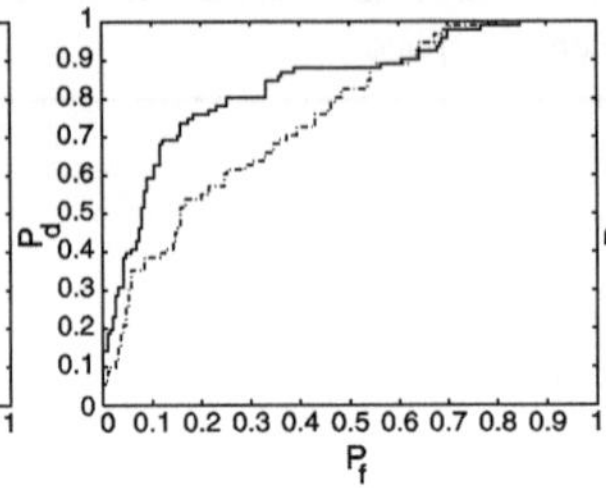


(b) ROC curve for $E_c$, $m_{ds}$

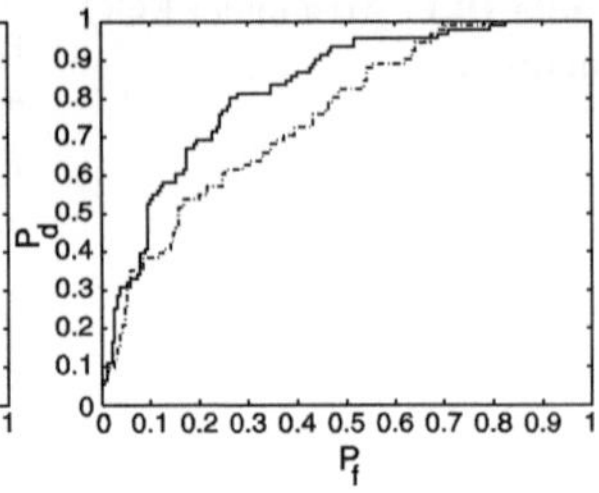


(c) ROC curve for $E_c$, $m_{lc1}$

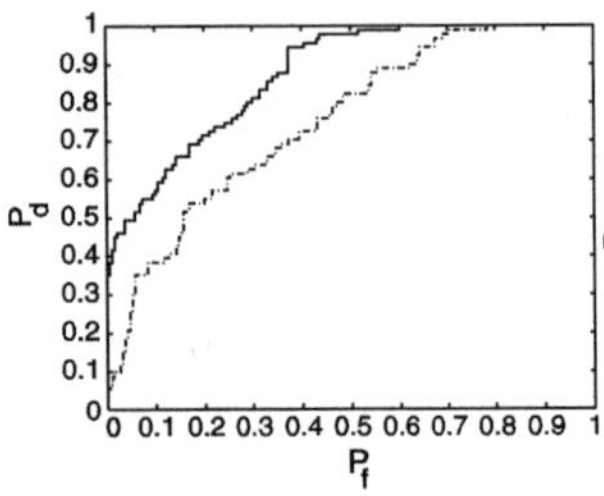


(d) ROC curve for $E_c$, $m_{lc2}$

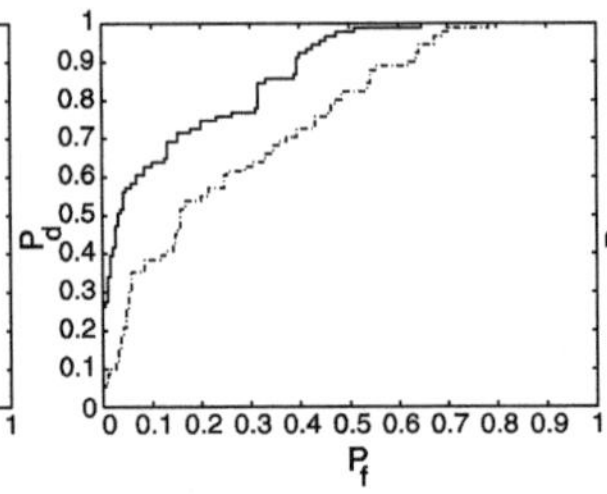


(e) ROC curve for $E_c$, $m_{lc3}$

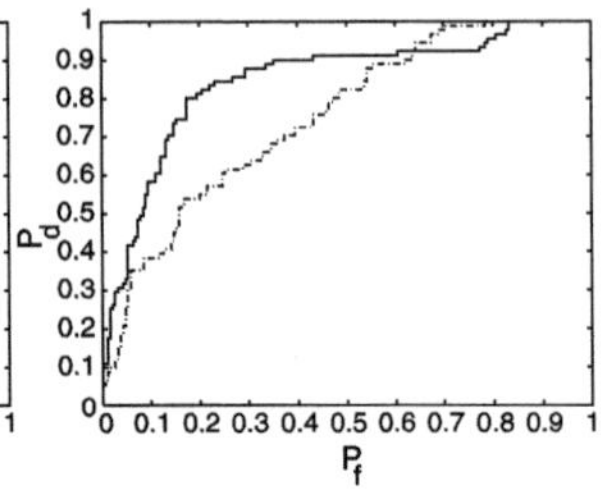


(f) ROC curve for $E_c$, $m_{fe}$

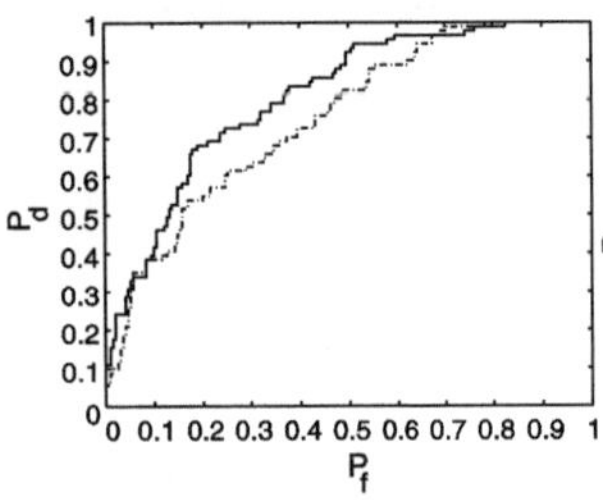


(g) ROC curve for $E_c$, $m_{ueg}$

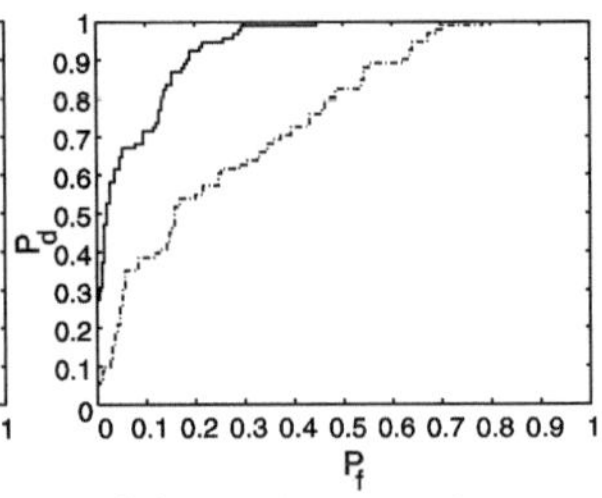


(h) ROC curve for $E_c$, $m_F$

**Fig. 10.2** ROC curves obtained by shifting the Bayesian decision boundary for residential region detection for the one class problem

## 10.2 Three-Class Problem

We next constructed a Bayes classifier, assuming equal priors, to label each image as rural, residential, or urban. We tested the classifier on 90 rural, 89 residential, and 91 urban images. We used five rural, four residential, and six urban representative samples for training. The training samples were chosen to cover a range of region types within each class.

Classification results using the structural features $\mu_c$, $E_c$ alone appear in Table 10.2. The overall correct classification rate in this space is 75.2%. We then conducted an exhaustive study of all possible feature spaces drawn from the full set of graph theoretical and structural measures (all singles, pairs, triples, etc.). Interestingly, the top three performing spaces are all two-dimensional, each consisting

**Table 10.2** Three class classification results in the $(\mu_c, E_c)$ space

| | Percentage classified as | | |
|---|---|---|---|
| | rural | residential | urban |
| 90 rural samples | 75.6 | 23.3 | 1.1 |
| 91 residential samples | 16.5 | 71.4 | 12.1 |
| 89 urban samples | 0.0 | 21.3 | 78.7 |

**Table 10.3** Three class classification results in the $(E_c, m_{ds})$ space

| | Percentage classified as | | |
|---|---|---|---|
| | rural | residential | urban |
| 90 rural samples | 97.8 | 0.0 | 2.2 |
| 91 residential samples | 13.2 | 78.0 | 8.8 |
| 89 urban samples | 0.0 | 27.0 | 73.0 |

**Table 10.4** Three class classification results in the $(E_c, m_F)$ space

| | Percentage classified as | | |
|---|---|---|---|
| | rural | residential | urban |
| 90 rural samples | 77.8 | 22.2 | 0.0 |
| 91 residential samples | 4.4 | 93.4 | 2.2 |
| 89 urban samples | 0.0 | 27.0 | 73.0 |

**Table 10.5** Three class classification results in the $(\mu_c, m_{ds})$ space

| | Percentage classified as | | |
|---|---|---|---|
| | rural | residential | urban |
| 90 rural samples | 85.6 | 13.3 | 1.1 |
| 91 residential samples | 6.6 | 82.4 | 11.0 |
| 89 urban samples | 0.0 | 27.0 | 73.0 |

of one structural feature and one graph-theoretical (or the fused) feature. They are $(E_c, m_{ds})$ (83.0% overall classification rate, Table 10.3); $(E_c, m_F)$ (81.5% overall classification rate, Table 10.4); and $(\mu_c, m_{ds})$ (80.4% overall classification rate, Table 10.5). It is not clear that these differences are statistically significant.

While the $(E_c, m_{ds})$ space provides the highest overall classification performance in these tests, the $(E_c, m_F)$ space is far better at detecting residential regions (93.4%), at the cost of greater false alarms from urban and rural areas. In our particular application, detecting residential areas, the $(E_c, m_{ds})$ space would therefore be preferred. For us, a false alarm is far less damaging than a miss. But for other applications, another space may prove superior.

Table 10.6 gives the difference ($\Delta$) between Tables 10.4 and 10.2 to highlight the performance differences offered by the fused graph theoretical measure. A neg-

**Table 10.6** Change in classification performance for the $(E_c, m_F)$ space relative to the $(\mu_c, E_c)$ space

| | Percentage classified as | | |
|---|---|---|---|
| | rural | residential | urban |
| 90 rural samples | 2.2 | −1.1 | −1.1 |
| 91 residential samples | −12.1 | 22.0 | −9.9 |
| 89 urban samples | 0.0 | 5.7 | −5.7 |

ative value in a diagonal entry, or a positive value in an off-diagonal entry, indicates a reduction in classification performance in moving from the "original" $(\mu_c, E_c)$ space to the $(m_E, m_F)$ space. The urban region detection rate decreases using the graph based measures, yet there is a clear **22.4%** improvement in residential region classification.

## 10.3 Summary of the Chapter

Our objective in this chapter was to detect residential regions which are difficult to discriminate. We first cast this as a one-class problem and obtained a clear and definite improvement by using graph-theoretical and structural features together over structural ones alone. This indicates that, for fine classification, structural features alone are inadequate. Besides the photometric structure in the scene, the spatial relationships of these structures, as captured by a graph and expressed in our measures here, are also needed.

Finally, we built a three-class classifier. For the three-class case, we obtained classification results superior to those reported anywhere over a similar number of test samples. Again, the results underscore the effectiveness of graph-theoretical measures for fine(r) classification. With these very high detection rates, our approach can be used to detect residential regions automatically.

## 10.4 Problems

**10.1** How do one- and three-class problems differ?

**10.2** What does the area below an ROC curve represent?

**10.3** Based on Tables 10.2, 10.3, and 10.4, which classes interfere more?

**10.4** Do graph-theoretical measures make a difference?

**10.5 (Open ended question)** Can there be other methods to extract the residential regions using graph-theoretical measures?

# Part V
# Building and Road Detection

# Chapter 11
# Review on Building and Road Detection

In this chapter, we briefly review the literature on building and road detection. We group previous studies as follows: building detection (alone), road detection (alone), and combined building and road detection. The last group of studies exploits the mutual information between buildings and road networks to improve detection.

## 11.1 Building Detection

Mayer [1] surveyed object detection systems from aerial images, focusing on building detection. In his excellent survey, he classified building detection systems based on their complexity (in data, building model, and system strategy). Following Mayer's format (in simplified form), we add missing and new papers published after his survey in Table 11.1. Besides the performance measures, Shufelt and McKeown [2] offer another survey on aerial building detection.

In Table 11.1, *data complexity* summarizes the resolution of the input image, relative location of the buildings, and the complexity of the scene. As for image types, some researchers used satellite images (such as Landsat, SPOT, and IRS) with resolutions in the 5 to 30 meter range. Others used aerial images with resolutions in the 0.3 to 1 meter range. Still others have used Digital Elevation Map (DEM) or Digital Surface Model (DSM) data. Finally, some have used Synthetic Aperture Radar (SAR) images. Each of these representations has its own benefits and shortcomings.

*Model complexity* characterizes the building model used. It may be a simple 2D rectangle, a 2D polynomial, or a 3D surface. The model should serve the system's application. If only detection is required, a simple rectangle as a building model may suffice. However, if a detailed site model is required, 3D surface models as well as polynomial representations are necessary.

*System strategy* captures the complexity of the system. Perceptual organization, Bayesian networks, and graph-theoretical methods are some of the approaches used in building detection systems.

Neither Mayer nor Shufelt and McKeown provided the classification performances of the systems they reviewed. Here, we fill this gap by providing reported

C. Ünsalan, K.L. Boyer, *Multispectral Satellite Image Understanding*, Advances in Computer Vision and Pattern Recognition, DOI 10.1007/978-0-85729-667-2_11, 

**Table 11.1** Building detection review, in Mayer's format

| Approach | Data | Model | Strategy |
|---|---|---|---|
| Huertas and Nevatia [4] | low-medium | simple | medium |
| Shufelt and McKeown [5] | low-medium | medium | medium |
| Krishnamachari and Chellappa [6] | low-medium | simple | medium |
| Maloof et al. [7] | low-medium | medium | medium |
| Brunn and Weidner [8] | high | medium | high |
| Zhang [9] | high | simple | simple |
| Kim and Muller [10] | low | simple | medium |
| Stassopoulou and Caelli [11] | low | medium | high |
| Gamba et al. [12] | high | medium | simple |
| Noronha and Nevatia [13] | low | simple | high |
| Fradkin et al. [14] | medium | high | medium |
| Krishnamoorthy et al. [15] | medium | medium | medium |
| Jaynes et al. [16] | medium | high | high |
| Peng et al. [17] | high | medium | medium |
| Katartzis and Sahli [18] | medium | high | high |
| Karantzalos and Paragios [19] | medium | medium | high |
| Sirmacek and Ünsalan [20] | medium | simple | high |
| Sirmacek and Ünsalan [21] | medium | simple | high |

classification performances in Table 11.2. In accordance with Lin and Nevatia [22], we cite probability of detection and branching factor. Probability of detection, $P_d$, is the percentage of the ground truth buildings detected. Branching factor, $B_f$, is the number of non-building objects (not pixels) labeled as buildings divided by the total number of objects labeled as buildings.

For the systems reported, $P_d$ varies from 41.5% to 100.0%, with $B_f$ in the 0.0% to 46.0% range. The performance depends on the resolution of the image, the density of buildings in the scene, and finally the size of the buildings to be detected. In most of these papers, the ground truth, such as the number and size of the buildings, has not been reported in detail. Also, the definitions of successful detection vary, ranging from simple detection of any part of a building [13], to complete delineation [23]. Any comparison of the numbers in Table 11.2 must be done in light of these *caveats*.

The size of the buildings to be detected affects the performance. For example, Kim and Muller [10] reported $P_d = 79.1\%$ for house detection over 12 houses, which is far less than the maximum $P_d$ obtained. Krishnamoorthy et al. [15] observed that most of their errors originate from house detection.

**Table 11.2** Building detection review, reported performances. Criteria for success vary; see text

| Approach | $P_d$ (%) | $B_f$ (%) |
|---|---|---|
| Shufelt and McKeown [5] | 49.8–86.6 | 12.8–32.2 |
| Maloof et al. [7] | 90.0–91.6 | |
| Lin and Nevatia [22] | 71.9 | 6.7 |
| Collins et al. [23] | 89.0 | 46.0 |
| Kim and Muller [10] | 76.3 | |
| Stassopoulou and Caelli [11] | 97.6 | 12.0 |
| Noronha and Nevatia [13] | 96.4 | 0.0 |
| Fradkin et al. [14] | 80.3 | 1.0 |
| Krishnamoorthy et al. [15] | 41.5–86.9 | |
| Jaynes et al. [16] | 99.5 | |
| Peng et al. [17] | 83.6 | |
| Katartzis and Sahli [18] | 100.0 | |
| Karantzalos and Paragios [19] | 98 | |
| Sirmacek and Ünsalan [20] | 88.4 | 14.4 |
| Sirmacek and Ünsalan [21] | 93.4 | 17.9 |

## 11.2 Road Detection

There is another excellent survey paper by Mayer et al. [3] on road detection in aerial images. As in the previous section, we add unreported papers in this survey following the same format (simplified) in Table 11.3. For each paper, we provide (if available) the data complexity, representation, and the resolution in meters per pixel.

As Table 11.3 shows, images with various complexities are used to detect road networks, including airborne, satellite, and SAR. Topology, parametric models, snakes, and semantic networks are the most popular representation methods. Resolution varies from 0.5 meters to 75 meters for these studies. In the low resolution (greater than 30 m/pixel) images only highways can be detected. There is also another review paper by Mena [40] summarizing road detection work till 2003.

In their survey, Mayer et al. did not provide road detection performances. As in building detection, we provide the reported road detection performances in terms of probability of detection $P_d$ and probability of false alarm $P_f$ in Table 11.4.

In Table 11.4, $P_d$ varies from 72.0% to 100.0%; $P_f$ varies from 1.0% to 10.6%. In these studies, the performance depends on the type of road to be detected, whether it is a highway with six lanes, a street in a city, or a road in a rural region. A method designed for a specific road type may not be useful for other types. Based on these, Table 11.4 gives a general idea about performance, but is not a conclusive test to pick the best method among them.

**Table 11.3** Road detection review

| Approach | Data complexity | Representation | Resolution (m) |
|---|---|---|---|
| Bajcsy and Tavakoli [24] | low | topology | 79 |
| Wang and Newkirk [25] | medium | semantic network | 10 |
| Fiset and Cavayas [26] | medium-high | topology, tracking | 30 |
| Netanyahu et al. [27] | medium | parametric | 1–3 |
| Tupin et al. [28] | high | semantic network | |
| Karathanassi et al. [29] | medium-high | topology, parametric | 6.25, 10 |
| Laptev et al. [30] | low | snakes | 0.5 |
| Jeon and Hong [31] | low | grouping, snakes | |
| Shi and Zhu [32] | high | topology | 1 |
| Stoica et al. [33] | medium | point processes | 5 |
| Mena and Malpica [34] | low | texture | 0.5 |
| Zhang and Couloigner [35] | low | texture | 1 |
| Hu et al. [36] | medium | tracking | 1 |
| Yang and Wang [37] | low | perceptual organization | 1 |
| Peng et al. [38] | high | variational model | 0.6 |
| Movaghati et al. [39] | low | particle, Kalman filtering | 1 |

**Table 11.4** Road detection review, reported performances. Criteria for success vary; see text

| Approach | $P_d$ (%) | $P_f$ (%) |
|---|---|---|
| Bajcsy and Tavakoli [24] | 85.0–100.0 | |
| Wang and Newkirk [25] | 87.7 | 10.6 |
| Fiset and Cavayas [26] | 79.3–81.3 | |
| Karathanassi et al. [29] | 92.0 | |
| Laptev et al. [30] | 72.0–84.0 | 1.0–5.0 |
| Jeon and Hong [31] | 92.2 | 1.6 |
| Shi and Zhu [32] | 91.5–92.1 | 1.6 |
| Stoica et al. [33] | | |
| Mena and Malpica [34] | 70.0 | |
| Zhang and Couloigner [35] | 84.0 | |
| Hu et al. [36] | 94.0 | |
| Yang and Wang [37] | 90.2 | 10.3 |
| Peng et al. [38] | 89.1 | |
| Movaghati et al. [39] | 98.0 | |

## 11.3 Combined Building and Road Detection

Some researchers have designed systems to detect both buildings and road networks simultaneously. Because buildings and road networks are not independent of each other, this strategy may improve the detection of both. Nagao et al. [41] and Nagao and Matsuyama [42] introduced such a system. Similarly, Hwang et al. [43] introduced a hypothesis generation and testing method to detect houses and road segments. In the next chapter, we also detect houses and street networks simultaneously.

## 11.4 Problems

**11.1** Where can road and building detection results be used?

**11.2** Why is detecting roads and buildings together more promising than detecting each alone?

**11.3** Besides using panchromatic images, what other sources of data can be used to detect buildings and road segments?

**11.4** (**Open ended question**) What are the possible future trends for road segment and building detection?

## References


1. H. Mayer, Comput. Vis. Image Underst. **74**, 138 (1999)
2. J.A. Shufelt, IEEE Trans. Pattern Anal. Mach. Intell. **21**(4), 311 (1999)
3. H. Mayer, A. Baumgartner, C. Steger, in *CV-Online* (1998)
4. A. Huertas, R. Nevatia, Comput. Vis. Graph. Image Process. **41**, 131 (1988)
5. R.B. Irvin, D.M. McKeown, IEEE Trans. Syst. Man Cybern. **19**(6), 1564 (1989)
6. S. Krishnamachari, R. Chellappa, IEEE Trans. Image Process. **5**(1), 164 (1996)
7. M.A. Maloof, P. Langley, S. Sage, T. Binford, in *Proceedings of the 1997 Image Understanding Workshop DARPA97* (1997), pp. 835–845
8. A. Brunn, U. Weidner, ISPRS J. Photogramm. Remote Sens. **53**, 296 (1998)
9. Y. Zhang, ISPRS J. Photogramm. Remote Sens. **54**, 50 (1999)
10. T. Kim, J.P. Muller, Image Vis. Comput. **17**(1), 3 (1999)
11. A. Stassopoulou, T. Caelli, Int. J. Pattern Recognit. Artif. Intell. **14**(6), 715 (2000)
12. P. Gamba, B. Houshmand, M. Saccani, IEEE Trans. Geosci. Remote Sens. **38**(1), 611 (2000)
13. S. Noronha, R. Nevatia, IEEE Trans. Pattern Anal. Mach. Intell. **23**(5), 501 (2001)
14. M. Fradkin, H. Maitre, M. Roux, Comput. Vis. Image Underst. **82**, 181 (2001)
15. P. Krishnamoorthy, K.L. Boyer, P.J. Flynn, in *Proceedings of the International Conference on Pattern Recognition*, vol. 1 (2002), pp. 159–163
16. C. Jaynes, E. Riseman, A. Hanson, Comput. Vis. Image Underst. **90**, 68 (2003)
17. J. Peng, D. Zhang, Y. Liu, Pattern Recognit. Lett. **26**, 587 (2005)
18. A. Katartzis, H. Sahli, IEEE Trans. Geosci. Remote Sens. **46**(1), 259 (2008)
19. K. Karantzalos, N. Paragios, IEEE Trans. Geosci. Remote Sens. **47**(1), 133 (2009)
20. B. Sirmacek, C. Ünsalan, IEEE Trans. Geosci. Remote Sens. **49**(4), 1156 (2009)
21. B. Sirmacek, C. Ünsalan, IEEE Trans. Geosci. Remote Sens. (2010, in press)
22. C. Lin, R. Nevatia, Comput. Vis. Image Underst. **72**(2), 101 (1998)
23. R.T. Collins, C.O. Jaynes, Y.Q. Cheng, X. Wang, F. Stolle, E.M. Riseman, A.R. Hanson, Comput. Vis. Image Underst. **72**, 143 (1998)

24. R. Bajcsy, M. Tavakoli, IEEE Trans. Syst. Man Cybern. **6**(9), 623 (1976)
25. F. Wang, R. Newkirk, IEEE Trans. Geosci. Remote Sens. **26**(5), 525 (1988)
26. R. Fiset, F. Cavayas, Int. J. Remote Sens. **18**(4), 991 (1997)
27. N.S. Netanyahu, V. Philomin, A. Rosenfeld, A.J. Stromberg, Pattern Recognit. **30**(10), 1673 (1997)
28. F. Tupin, H. Maitre, J.F. Mangin, J.M. Nicolas, E. Pechersky, IEEE Trans. Geosci. Remote Sens. **36**(2), 434 (1998)
29. V. Karathanassi, C. Iossifidis, D. Rokos, Int. J. Remote Sens. **20**(1), 153 (1999)
30. I. Laptev, H. Mayer, T. Lindeberg, W. Eckstein, C. Steger, A. Baumgartner, Mach. Vis. Appl. **12**, 23 (2000)
31. B.K. Jeon, J.H. Jang, K.S. Hong, IEEE Trans. Geosci. Remote Sens. **40**(1), 22 (2002)
32. W. Shi, C. Zhu, IEEE Trans. Geosci. Remote Sens. **40**(2), 511 (2002)
33. R. Stoica, X. Descombes, J. Zerubia, Int. J. Comput. Vis. **57**, 121 (2004)
34. J.B. Mena, J.A. Malpica, Pattern Recognit. Lett. **26**, 1201 (2005)
35. O. Zhang, I. Couloigner, Pattern Recognit. Lett. **27**, 937 (2006)
36. J. Hu, A. Razdan, J.C. Femiani, M. Cui, P. Wonka, IEEE Trans. Geosci. Remote Sens. **45**(12), 4144 (2007)
37. J. Yang, R.S. Wang, Int. J. Remote Sens. **28**, 4653 (2007)
38. T. Peng, I.H. Jermyn, V. Prinet, J. Zerubia, IEEE J. Sel. Top. Appl. Earth Obs. Remote Sens. **1**, 139 (2008)
39. S. Movaghati, A. Moghaddamjoo, A. Tavakoli, IEEE Trans. Geosci. Remote Sens. **48**(7), 2807 (2010)
40. J.B. Mena, Pattern Recognit. Lett. **24**, 3037 (2003)
41. M. Nagao, T. Matsuyama, Y. Ikeda, Comput. Vis. Graph. Image Process. **10**, 195 (1979)
42. M. Nagao, T. Matsuyama, *A Structural Analysis of Complex Aerial Photographs* (Plenum, New York,, 1980)
43. V.S. Hwang, L.S. Davis, T. Matsuyama, Comput. Vis. Graph. Image Process. **36**, 321 (1986)

# Chapter 12
# House and Street Network Detection in Residential Regions

In the previous chapters, we introduced several methods to detect residential regions starting from land use classification. In this chapter, we introduce a novel subsystem (of our multispectral satellite image understanding system) to detect houses and the street network in residential regions. Detecting houses is far more challenging than detecting larger buildings for several reasons. First, their relatively small size (in onemeter resolution Ikonos images) makes their detection difficult. Second, occlusion by nearby trees is common. Third, in some neighborhoods, houses may come in fairly complex shapes. Analogous problems (small cross-section, overhanging trees, and winding curves) present challenges for street detection in residential regions.[1]

Our house and street network detection subsystem comprises four main parts. We first introduce measures on multispectral images to detect regions of possible human activity. On these measures, we introduce a variation of the k-means clustering algorithm to extract possible houses and the street network by combining both spatial and spectral features. This combination of information improves the final clustering results. From clustering, we obtain a binary image containing possible street network fragments and houses. We then decompose this binary image using a balloon algorithm based on binary mathematical morphology. Having obtained the decomposition, we represent them in a graph theoretical framework. Balloons serve as vertices and their neighborhood information is encoded as edges in the graph. The street network is extracted from the graph by using the unary and binary constrains. The remaining vertices (balloons) are assigned as possible houses in the region. Next, we explore each part in detail.

## 12.1 Using Multispectral Information

Most prior work on building and road detection uses either panchromatic images or DSM data. Besides panchromatic images, we have additional multispectral in-

[1]The figures in this chapter are obtained from our previous work [3]. Here, they appear with the kind permission of Elsevier.

C. Ünsalan, K.L. Boyer, *Multispectral Satellite Image Understanding*,
Advances in Computer Vision and Pattern Recognition,
DOI 10.1007/978-0-85729-667-2_12, © Springer-Verlag London Limited 2011

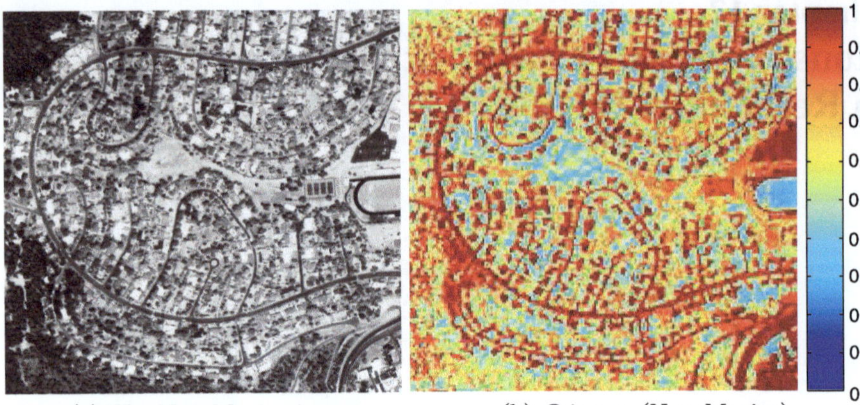


(a) The New Mexico image (b) $\Omega$ image (New Mexico)

**Fig. 12.1** $\Omega$ example

formation, but no surface or site models. We use this information to infer cultural activity (such as houses and street networks) and water (such as lakes).

## *12.1.1 A Derived Index to Detect Human Activity*

We can use vegetation indices, given in Sect. 3.3, to detect human activity in multispectral images. The *NDVI* and its linearized version $\theta$ respond with a low value (around 0) to rocks, stones, and their derivatives [1]. These materials are used extensively for building and road construction, thereby indicating possible human activity (actual rock outcroppings in residential regions are rarely large enough to be seen in Ikonos data). Also, $\theta$ achieves high values in areas of significant vegetation density, while returning negative values for cloud, shadow, and snow. Therefore, to measure possible building or street pixels in residential regions, we use the following derived index:

$$\Omega = 1 - |\theta| \tag{12.1}$$

$\Omega$ is normalized between 0 (suggesting low human activity) and 1 (usually meaning high human activity). Any rock outcroppings that may be detected in residential areas are easily eliminated in subsequent processing based on their irregular footprint.

To illustrate $\Omega$, we use a part of the New Mexico image given in Fig. 12.1(a). This neighborhood shows characteristics typical of a mature residential region. Each house has its own garden; they are well-spaced; and there are mature trees nearby. If we consider Fig. 12.1(a), we can appreciate the challenging nature of the house and street detection problem in residential regions. Even human observers find it difficult to decide whether there are houses in some parts of the image without using context, such as partial knowledge of the street layout, driveways, and so on. We give the corresponding color-coded (blue corresponds to the lowest and red corresponds to the highest values) $\Omega$ image in Fig. 12.1(b). Houses and the street network are represented by high index values (in red) in this image.

### *12.1.2 Using a Shadow–Water Index to Eliminate Lakes*

While high $\Omega$ values indicate possible house and street network pixels, they may also correspond to water regions such as rivers, lakes, or ponds. Because such features can appear in residential regions, we eliminate them using a method we introduce in this section.

In the Ikonos spectrum, water shows an increasing response curve until the blue band. It reaches a maximum in this region and then decreases monotonically to the near-infrared [1]. So, a representative shadow water index should be composed of high blue values first. Ideally, it should also consider the green and red bands, but the green band also responds strongly to vegetation and this impairs the shadow or water observation. Hence, the index should at least include blue and red bands. Under the light of these observations, we use the shadow–water index, $\gamma_2$, introduced in Chap. 4.

Because this index is unable to discriminate water from shadows, we use $\theta$ and $\gamma_2$ in conjunction to distinguish the water regions. To do so, we first obtain a binary image from the $\gamma_2$ image. Our tests indicate that pixels having $\gamma_2$ values higher than 0.3 are possible shadows or water regions. Therefore, we threshold the $\gamma_2$ image at this value. We then use connected components analysis to extract eight-connected regions in this binary image. We eliminate regions smaller than 75 pixels, considering them to be insignificant (neither a significant body of water nor a shadow region that will impact subsequent processing). Because water gives a strong $\gamma_2$ response, but a low $\theta$ response, while shadows generally cover at least some vegetation, we label a region (as segmented from the $\Omega$ image, details below) as water if its $\theta$ median is less than 0.2 and its $\gamma_2$ median is greater than 0.3.

Water detection was applied to the 44 images at hand, and all bodies of water covering more than 75 pixels (19 in all) were detected. There were no false positives. As an example, Fig. 12.2 shows an Indiana image containing a lake in the middle of the scene. We give its panchromatic image, color-coded (blue corresponds to the lowest value, red corresponds to the highest value) $\gamma_2$ image, and the water region detected. The lake is correctly detected in the middle of the residential region.

## 12.2 Segmenting the $\Omega$ Image

Although $\Omega$ indicates possible human activity, we need a binary image for subsequent processing. In this binary image, possible human activity will be labeled as foreground. This section introduces an extension of the k-means clustering algorithm that exploits spatial coherence via connected components for segmentation.

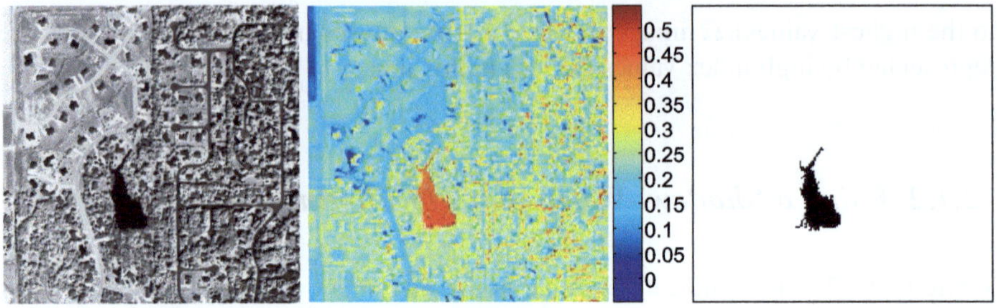


(a) The Indiana image (b) $\gamma_2$ image (Indiana) (c) The lake detected

**Fig. 12.2** Detecting water regions

---

**Algorithm 1**: K-means clustering

---

Begin
$n$ = number of samples (pixels)
$c$ = number of classes
$\mu_i$ = mean vector for class $i$
Initialize $n, c, \mu_1, \mu_2, \ldots, \mu_c$
**repeat**
  classify $n$ samples according to nearest $\mu_i$
  recompute $\mu_i$
**until** no change in $\mu_i$
return $\mu_1, \mu_2, \ldots, \mu_c$
End.

---

## *12.2.1 K-means Clustering with Spatial Coherence*

K-means clustering (KMC) is a standard iterative technique in pattern recognition to extract natural clusters in the data [2, 4]. Standard k-means appears as Algorithm 1. In standard k-means clustering, the Euclidean distance is used in the classification step. Jain et al. [5] suggested using the Mahalanobis distance instead. No spatial information is used in either case.

As the zoomed (New Mexico) $\Omega$ image (Fig. 12.3) illustrates, houses and street segments are connected via driveways. The street network is also connected. To use this information, we extend the k-means clustering algorithm by introducing spatial information derived from connected components analysis. We call this algorithm *k-means clustering with spatial coherence* (KMC-SC).

Our application of k-means clustering with spatial coherence is as follows. On the $\Omega$ image, we define an initial segmentation threshold $t_s$ (we show below that this value is not critical). Pixels for which $\Omega > t_s$ are the initial object (foreground) pixels (possible human activity); the rest are the initial background pixels. These serve as input to the class-conditional sample mean and covariance calculations. K-means clustering using the Mahalanobis distance is then applied to the data to update the

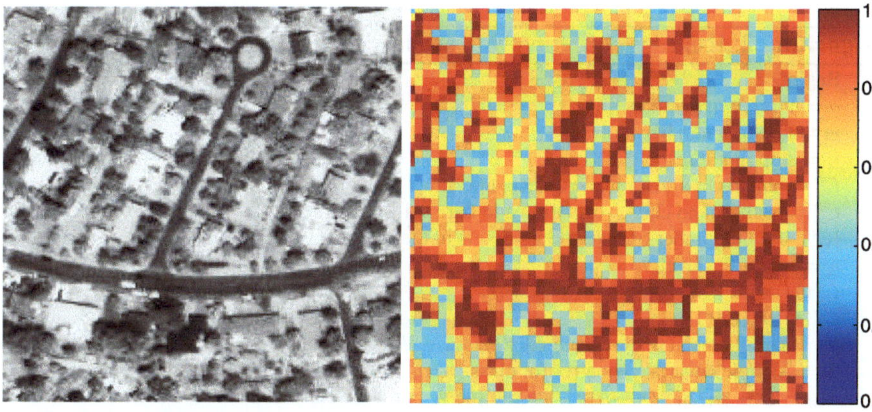


(a) Zoomed New Mexico image (b) Zoomed $\Omega$ image (New Mexico)

**Fig. 12.3** $\Omega$ example, zoomed

object and background hypotheses. Here, the use of spatial coherence begins. The current object pixel hypotheses are decomposed into eight-connected sets. Those sets intersecting with the initial set of object pixels (at the beginning of this iteration) form the final object segment for this iteration. This maintains object class coherence throughout clustering; no new, isolated object segments can be created. We apply the spatial coherence constraint only to the foreground. A new iteration starts by updating the sample mean and covariance calculations. The iteration terminates when the absolute value of the difference between the current and previous object (total) pixel counts falls below a threshold. The details appear in Algorithm 2.

We present iteration steps for this algorithm on the Indiana image in Fig. 12.4. After initial thresholding, some parts of the object segments have been obtained. As the algorithm iterates, it adds remaining object pixels to the initial object segment. Finally, most of the object segments have been labeled correctly. Importantly, pixels added early in the process present greater $\Omega$ values than those added later.

We demonstrate the algorithm on four images from different regions. For each case, we give the panchromatic image and the binary segment ($C^o$) in Fig. 12.5, 12.6. On all test images, our clustering method was able to converge in a reasonable number of iterations (approximately 10 to 12 iterations on average). These experiments show that our clustering method works fairly well on different residential regions and varying environmental conditions including changes in vegetation cover and soil type.

### *12.2.2 Comparison with Other Methods*

We will now compare our modified clustering method with standard k-means clustering, Otsu's method [6], and region growing [7, 8]. We also tested fuzzy k-means

**Algorithm 2**: K-means clustering with spatial coherence

Begin
$n$ = number of samples (pixels)
$m$ = iteration number; $m := 1$
$t_e$ = stopping threshold
$C^o$ = object class, man made objects like street networks and houses (image space)
$C^b$ = background class, vegetation and bare soil (image space)
$|C^o|$ = total number of object class pixels in the image
$t_s$ = initial $C^o$ level, $t_s := 0.8$
$C^o(1) := \Omega > t_s$
$C^b(1) := (C^o(1))^c$
$\mu_\mathrm{i}$ = sample mean feature vector for class $i \in \{b, o\}$
$\mathbf{\Sigma}_\mathrm{i}$ = sample covariance matrix for class $i \in \{b, o\}$ (feature space)
$D_M = (\mathbf{x} - \mu_\mathrm{i})\mathbf{\Sigma}_\mathrm{i}^{-1}(\mathbf{x} - \mu_\mathrm{i})^T$, $\mathbf{x}$ a sample vector to be classified
compute $\mu_\mathrm{i}$, $\mathbf{\Sigma}_\mathrm{i}$, $i \in \{b, o\}$
**repeat**
  $m := m + 1$
  classify $n$ samples according to nearest class via Mahalanobis distance $D_M$
  find $C^o(m)$ in image space
  find $k$ connected components of $C^o(m)$, $C^o_j(m)$ $j = 1, 2, \ldots, k$
  $C^o(m) := \bigcup_{j\in\{1,2,\ldots,k\}} C^o_j(m)$ such that $C^o(m-1) \bigcap C^o_n(m) \neq \emptyset$
  $C^b(m) := (C^o(m))^c$
  recompute $\mu_i$, $\Sigma_i$, $i \in \{b, o\}$
**until** $||C^o(m)| - |C^o(m-1)|| < t_e$
return $C^o(m)$
End.

clustering [2] but the performance was similar to k-means clustering; therefore, we omit it here.

Otsu formulates the optimum threshold by maximizing the between-class variance assuming a bimodal histogram such that one peak corresponds to the object and the other to the background. This method depends on the histogram only. Although it is extremely fast, the results become less reliable if the bimodal histogram assumption is not satisfied.

In region growing, initial seed points are selected by some appropriate process. Then, nearby points are added to these seed points such that they do not exceed a threshold in variance or another homogeneity criterion. Here, we obtained the seed points by thresholding, and for homogeneity we used the variance threshold. One shortcoming of region growing is its heavy dependence on seed points. Also, as the region area grows, the incursion of false pixels to clusters becomes more probable [8].

We compare these four methods in Fig. 12.7. Here, we use four different residential regions from Indiana, Maryland, New Mexico, and South Dakota. Segmentation by standard k-means clustering (KMC) performed worst of all. We see the effects of heavy dependence on the seed points in region growing on the Indiana and Maryland images. The method cannot recover regions having no seed points. Finally, KMC-SC performed well on all four images. We also quantify these results in Table 12.1.

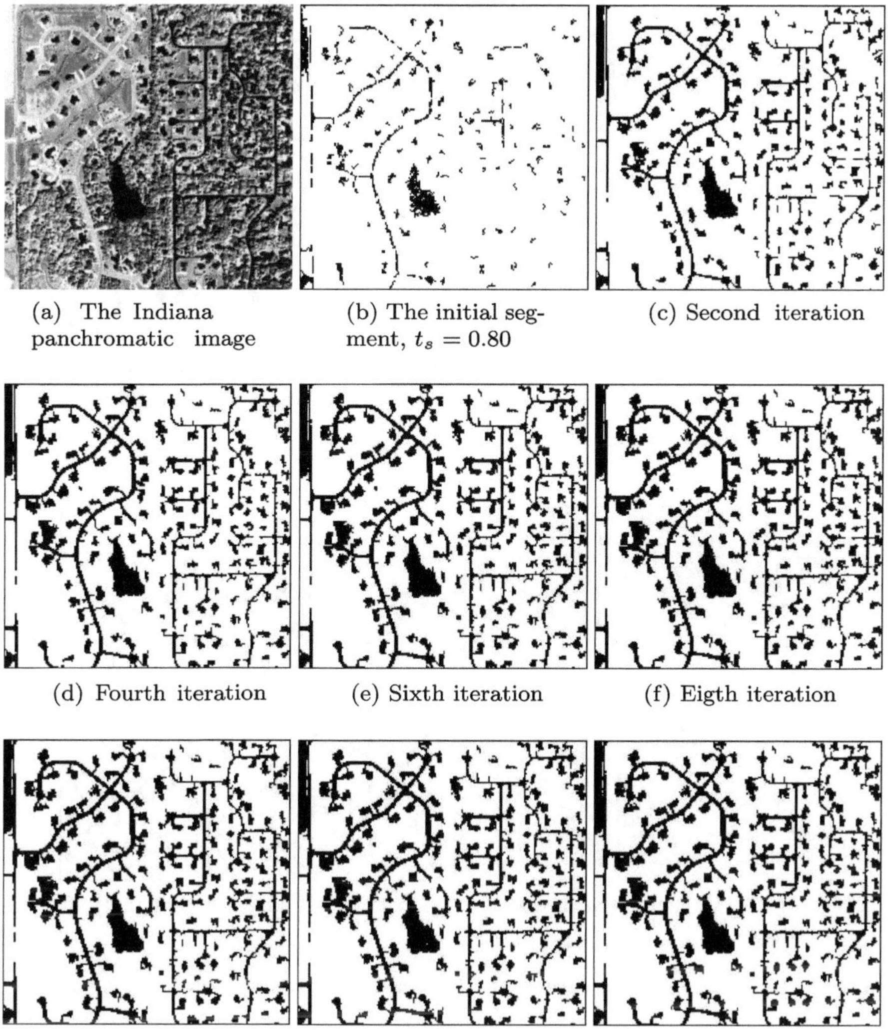

(a) The Indiana panchromatic image
(b) The initial segment, $t_s = 0.80$
(c) Second iteration
(d) Fourth iteration
(e) Sixth iteration
(f) Eigth iteration
(g) Tenth iteration
(h) Twelfth iteration
(i) The final segment ($C^o$)

**Fig. 12.4** Sample iterations on the Indiana image by Algorithm 2, KMC-SC

As Table 12.1 shows, KMC-SC has the highest $P_d$ over three images. It falls behind Otsu's method only in the New Mexico image. Otsu's method, however, produces a significant number of disconnected segments because it invokes no spatial coherence constraint. Region growing has the lowest $P_f$ for three images except South Dakota. However, for these three images, it has $P_d$ values far below KMC with spatial coherence. Overall, KMC-SC is the best segmentation method among these four for our application.

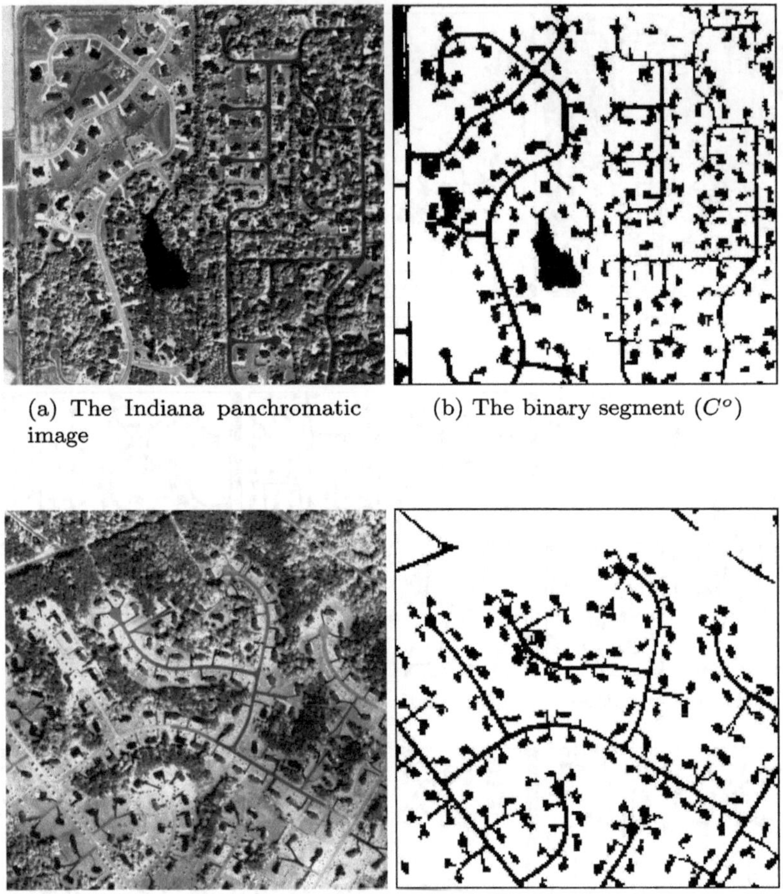

(a) The Indiana panchromatic image

(b) The binary segment ($C^o$)

(c) The Maryland panchromatic image

(d) The binary segment ($C^o$)

**Fig. 12.5** Test images and binary segments $C^o$ returned by Algorithm 2, KMC-SC, Part I

## *12.2.3 Dependence on Initial Conditions*

The only free parameter in KMC-SC is the initial segmentation threshold $t_s$. This also serves as the seed point extraction step in region growing. Here, we test the stability of our method with respect to this initial threshold value. We consider three different initial segmentation thresholds ranging from $t_s$ very large to very small. We give the initial and the final segments in Fig. 12.8. We also provide the total pixel count in the segment versus the iteration number, showing that convergence is nearly exponential.

The results show that the selection of $t_s$ is not highly critical. If the segmentation threshold is high (fewer seed points), the method is able to recover most of the missing parts. If the segmentation threshold is low (seed points cover more area

(a) The New Mexico panchromatic image

(b) The binary segment ($C^o$)

(c) The South Dakota panchromatic image

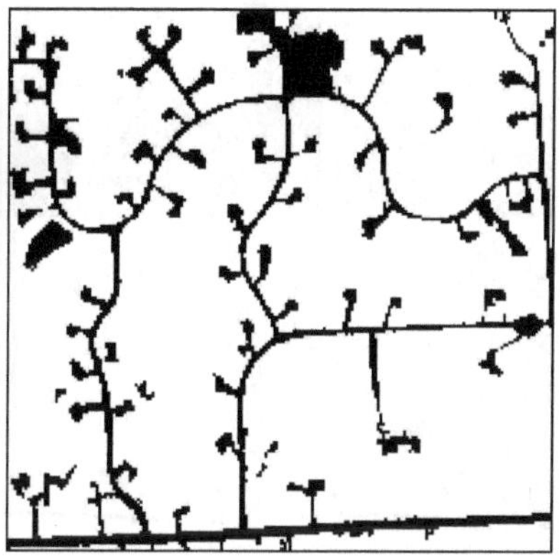

(d) The binary segment ($C^o$)

**Fig. 12.6** Test images and binary segments $C^o$ returned by Algorithm 2, KMC-SC, Part II

**Table 12.1** Comparison of performances, in percentages

| | Indiana | | Maryland | | New Mexico | | South Dakota | |
|---|---|---|---|---|---|---|---|---|
| | $P_d$ | $P_f$ | $P_d$ | $P_f$ | $P_d$ | $P_f$ | $P_d$ | $P_f$ |
| KMC | 78.0 | 49.0 | 86.3 | 38.7 | 20.2 | 56.9 | 96.2 | 34.1 |
| Otsu's method | 91.9 | 6.4 | 85.3 | 1.6 | **96.7** | 36.6 | 98.2 | 7.9 |
| Region growing | 26.8 | **2.1** | 25.4 | **0.2** | 74.0 | **6.7** | 96.9 | 6.6 |
| KMC-SC | **94.4** | 5.6 | **96.4** | 1.0 | 95.3 | 13.2 | **98.7** | **3.3** |

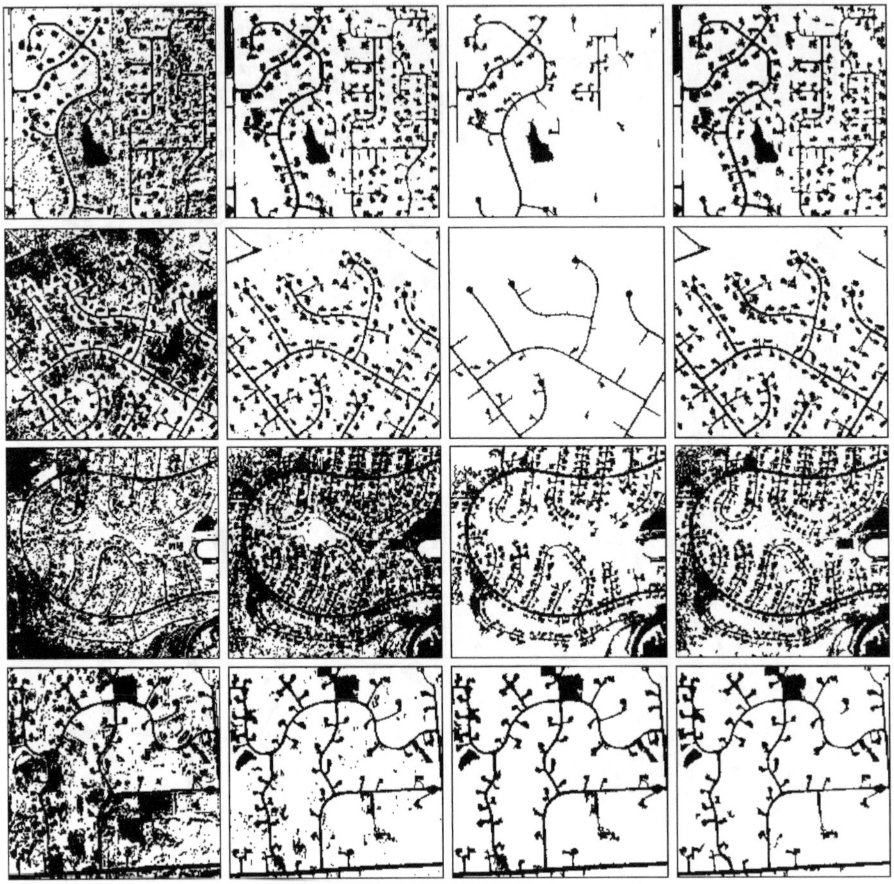

**Fig. 12.7** Comparison of segmentation algorithms. *First row*: Indiana image; *second row*: Maryland image; *third row*: New Mexico image; *fourth row*: South Dakota image. *First column*: KMC; *second column*: Otsu's method; *third column*: region growing; *fourth column*: KMC-SC

than the actual shape), the method is able to eliminate most extra parts. Of course, if a reasonable threshold is selected we get the best performance. The threshold can also be calculated by another algorithm (such as Otsu's method); however, we found no significant gain in doing so. Since the $\Omega$ image is scaled, a fixed threshold for all images suffices; we set it to 0.8.

The case of Fig. 12.8(h) is interesting in that the lake was eliminated by KMC-SC, prior to the explicit water elimination step described above. This was simply the outcome of KMC-SC for this case. Because water regions present intermediate $\theta$ values, the cluster which they will appear in is unpredictable. This observation motivated the water elimination technique presented above.

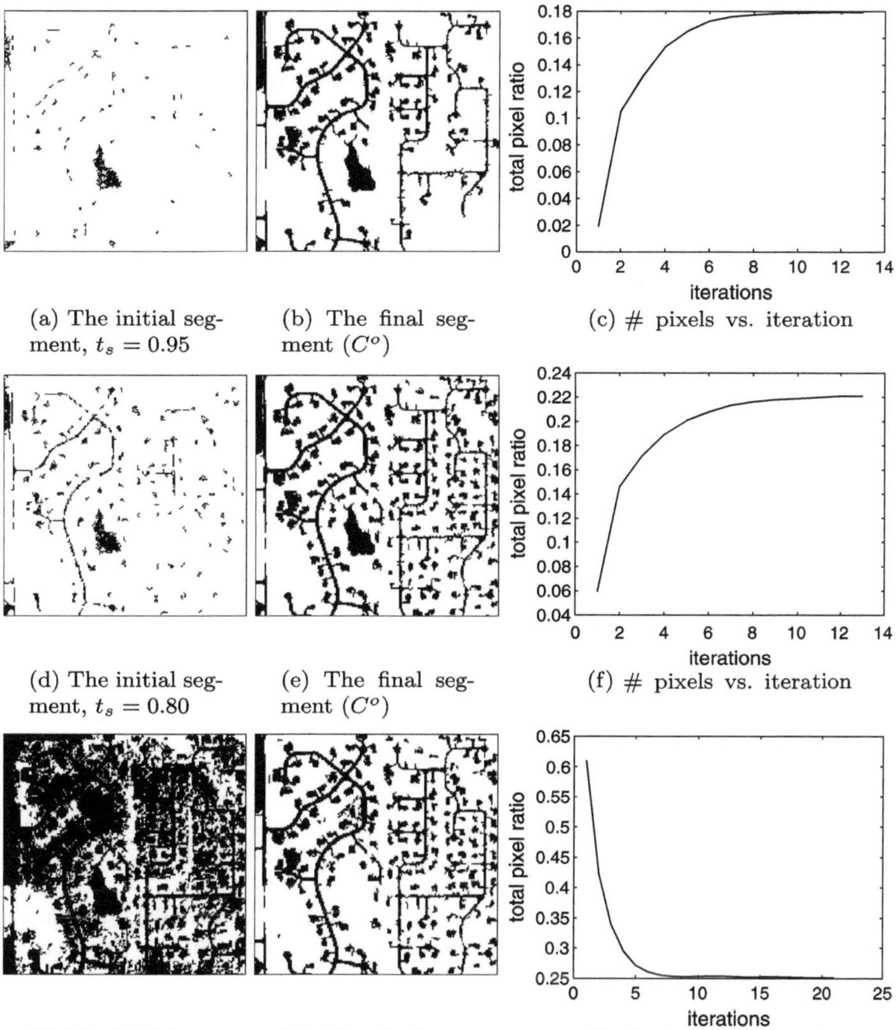


(a) The initial segment, $t_s = 0.95$
(b) The final segment ($C^o$)
(c) # pixels vs. iteration
(d) The initial segment, $t_s = 0.80$
(e) The final segment ($C^o$)
(f) # pixels vs. iteration
(g) The initial segment, $t_s = 0.40$
(h) The final segment ($C^o$)
(i) # pixels vs. iteration

**Fig. 12.8** The Indiana image, segmentation results with different $t_s$ values

## 12.3 Shape Decomposition of the Foreground: The Binary Balloon Algorithm

We next decompose the binary segment $C^o$ (representing possible houses and the street network) into subsets such that each either represents a structure (a house or a street segment), or a combination represents such a structure. From this point on, $C^o$ represents a binary segment without lakes and water regions. We discussed how to discard these regions in Sect. 12.1.2.

For decomposition, we introduce a binary balloon algorithm. Balloons are deformable models that are fitted to a target object in an image using external and internal constraints [9]. External constraints are provided by image (photometric) gradients, while the internal constraints are smoothness and bending forces designed to limit the balloon's shape complexity. Balloon algorithms have been used extensively for shape fitting in computer vision.

The extent of the objects in $C^o$ motivates us to introduce a new balloon algorithm tailored to our domain. We construct the binary balloons by set morphology operations. Mathematical morphology is a well-known tool to extract regions based on their shapes [8, 10]. We now provide a very brief review of elementary mathematical morphology. Mathematical morphology (more properly, Minkowski algebra) is based on logical operations over, in our case, sets in $Z^2$. We need three basic morphology set operations: translation, dilation, and erosion. We consider each in turn. Let the points (vectors) be $\mathbf{p} = (p_1, p_2)$, $\mathbf{a} = (a_1, a_2)$, $\mathbf{b} = (b_1, b_2)$. Let $A$ be a set of points, not necessarily connected, in the plane.

The *translation* of set $A$ by $\mathbf{p}$ is:

$$(A)_p = \{c | c = a + p, a \in A\}. \tag{12.2}$$

The *dilation* of set $A$ by set $B$ is:

$$A \oplus B = \bigcup_{b \in B} (A)_b. \tag{12.3}$$

The *erosion* of set $A$ by set $B$ is:

$$A \ominus B = \bigcup_{b \in B} (A)_{-b}. \tag{12.4}$$

The set $B$ as used in (12.3), and (12.4) is commonly referred to as the "structuring element". It is a sort of template for the analysis of the shape of the set $A$.

### 12.3.1 Constructing Initial Balloons

Our primary goal in decomposing the binary image $C^o$ is to identify its *elongated subsets* such that they represent possible street segments (non-elongated segments will represent other structures, including houses). For this purpose, we use lines as initial balloons in $C^o$. We extract lines by applying erosion and dilation operations, respectively. We define four linear structuring elements (one could use more, but we found no advantage in doing so): horizontal, vertical, the two diagonals. If we take $x$ and $y$ as coordinates in the image space, horizontal corresponds to a constant $y$, vertical corresponds to a constant $x$, diagonal-1 corresponds to a line segment $y = x + b$, $b$ being the $y$-intercept, and diagonal-2 corresponds to a line segment $y = -x + c$, $c$ being the $y$-intercept.

**Algorithm 3**: Algorithm to construct initial balloons

```
Begin
l = length of the structuring element
M = binary image (in our case obtained from segmentation)
S(l) = structuring element
l := maximum length
while l > 0 do
   E := M ⊖ S(l)
   if E ≠ ∅ then
      D := E ⊕ S(l)
      Call the binary balloon algorithm (Algorithm 4) [input = D, M; output = B]
      M := M \ B
   else
      l := l − 1
   end if
end while
End.
```

To extract initial balloons for each structuring element, we apply Algorithm 3. The procedure is briefly described as follows. We set the length of the structuring element to the maximum image size to detect the longest initial balloon possible. We then decrease the length of the structuring element until we detect an instance of it in the image. This structuring element is taken as an initial balloon. Then, we expand the initial balloon to construct the corresponding binary balloon and eliminate its pixels from the image via set subtraction. By decreasing the size of the structuring element, we continue to detect initial balloons and expand the corresponding binary balloons. Since we discard the points in each binary balloon before extracting the next, the final set of balloons will be disjoint.

### 12.3.2 *Inflating the Binary Balloons*

Once we obtain an initial balloon, we construct the corresponding binary balloon. We expand the initial balloon to fit into a region similar to standard balloon algorithms. However, unlike other algorithms, ours does not use a parametric form for the balloon. Instead, we expand each balloon by adding neighboring points satisfying a *smoothness condition*. This is a unique feature of our binary balloon algorithm; standard balloon algorithms perform poorly in our domain owing to the extreme aspect ratios of streets and roads, and small house sizes.

We start with a binary balloon obtained from Algorithm 3. The outer boundary points (eight-connected neighbors in the object class) of the initial balloon are labeled as candidate points. If there are no such foreground pixels, the balloon expands no more. Therefore, the balloon can never expand outside the object region. We group the candidate pixels into eight-connected candidate sets by connected components analysis. To ensure a smooth boundary, each added candidate set should be

**Algorithm 4**: The binary balloon algorithm

```
Begin
D = initial balloon obtained by Algorithm 3
M = binary image modified by Algorithm 3
t_o = overlap threshold
m = iteration number; m := 1
B = binary balloon (a region in image space)
P = three by three pixel square set
|B| = total number of "on" pixels in set B
B := D
C(0) := D
while A ≠ ∅ do
   C(m) := (C(m − 1) ⊕ P) ∩ M
   A := C(m) \ C(m − 1)
   A_n := connected components of A; n = 1, 2, ..., k
   for n = 1 to k do
      if |B ∩ (A_n ⊕ P)| > t_o then
         B := B ∪ A_n
      end if
   end for
   m := m + 1
end while
return B
End.
```

similar to the balloon locally. That is, we want the added set to be elongated along the same axis. This will be our internal constraint. Since all pixels in each candidate set are eight-adjacent to the current balloon, larger sets are more consistent in shape (orientated elongation) with the current balloon than are the smaller sets. Therefore, we define a size threshold $t_o$ to control the boundary smoothness; this is set to a fixed fraction of the number of pixels in the initial balloon. If we keep this threshold high, the decomposition of $C^o$ will consist of more elongated balloons with smoother boundaries; if we keep it low, fewer, less elongated balloons with rougher boundaries will result. Only if the total number of adjoining pixels exceeds $t_o$, which we set $t_o$ equal to one-half the number of pixels in the initial balloon, do we add the candidate set to the balloon. We update the balloon, and iterate until the balloon expands no more.

Initial additions will be lines, since we start with a line. As the balloon starts to take the shape of the object, curved lines can be added. At every step, the size threshold will force the shape to remain smooth. The binary balloon construction algorithm appears in Algorithm 4.

To clarify these steps, we demonstrate them on a small example in Fig. 12.9. Here, dashed blue lines represent candidate points, and red points represent the extracted balloon for each iteration. We start with the initial horizontal balloon (labeled red) in Fig. 12.9(b). Two candidate sets (labeled in green) exceed $t_o$ (separately) at the first iteration, and therefore will be added to the balloon. Two candidate sets (labeled yellow) could not exceed $t_o$ at the second iteration. At this point, there are

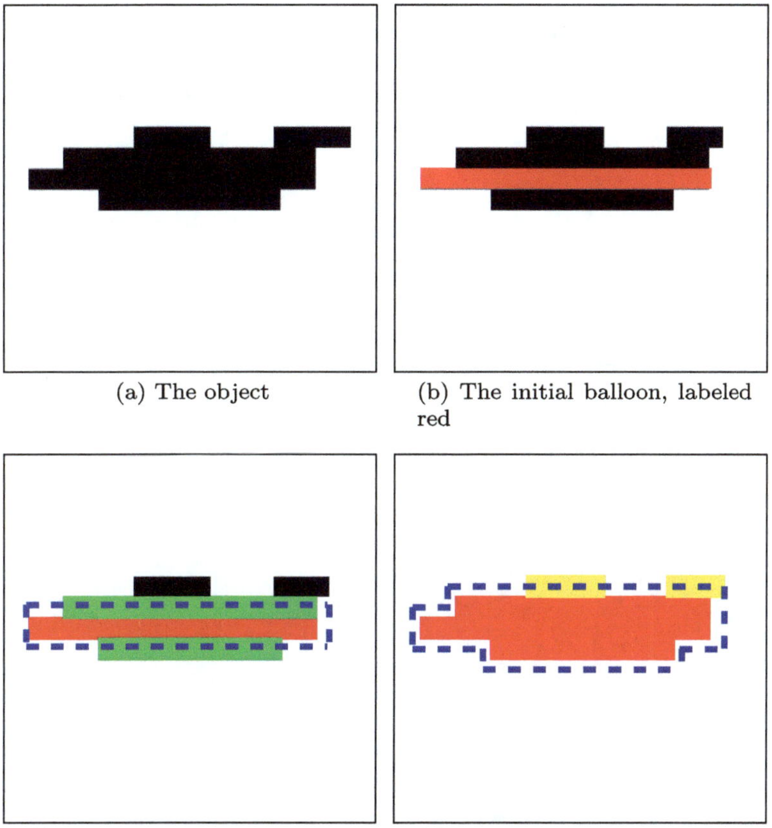

(a) The object

(b) The initial balloon, labeled red

(c) The first iteration, blue - candidate points, green - accepted candidate sets

(d) The second iteration, yellow - rejected candidate sets, red - the final balloon

**Fig. 12.9** Extracting the binary balloon, an example

no more candidate object pixels and the iteration stops. The final balloon extracted is labeled in red in Fig. 12.9(d).

Next, we demonstrate binary balloon extraction on the South Dakota image. To obtain the decomposition for $C^o$ in this image, we obtain all the binary balloons in $C^o$ with the four structuring elements separately as illustrated in Fig. 12.10. For demonstration purposes, we applied a greedy graph-coloring algorithm in all these figures, such that no two neighboring balloons have the same color [11]. In this figure, gray levels have no meaning except as labels to distinguish one balloon from another.

In Fig. 12.10(a), the decomposition of $C^o$ into horizontal balloons is shown. In this decomposition, there are around 200 horizontal non-intersecting balloons covering $C^o$. Horizontal $C^o$ sections are represented by a few large balloons; curved sections are represented by many small balloons. This is also the case for vertical and diagonal decompositions with respect to their initial balloons. All balloons have

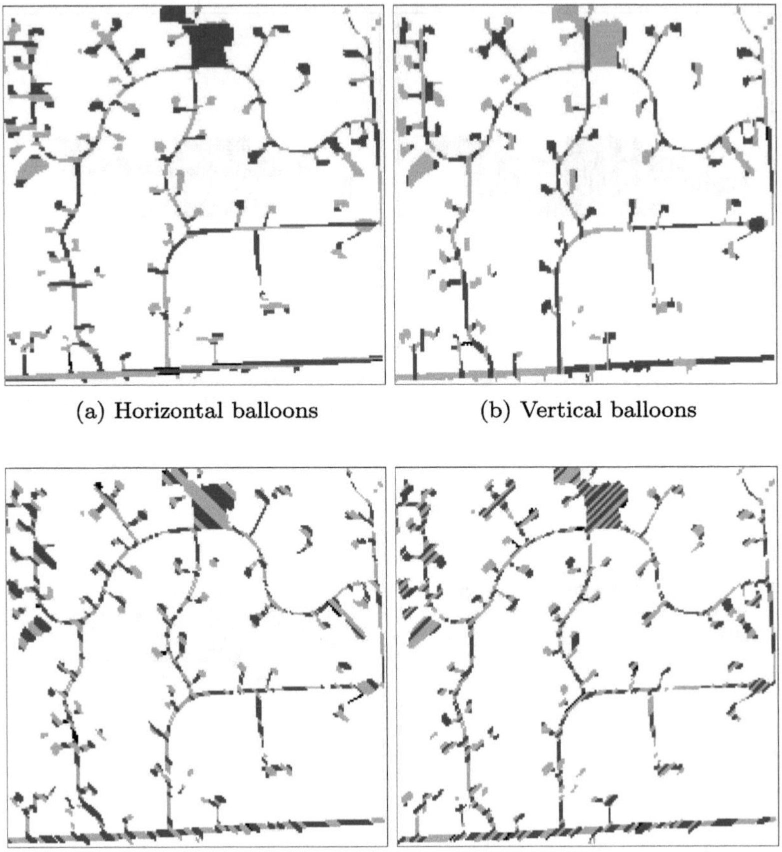

(a) Horizontal balloons (b) Vertical balloons

(c) Diagonal-1 balloons (d) Diagonal-2 balloons

**Fig. 12.10** The South Dakota image, decomposition of $C^o$

smooth boundaries because of the size threshold (internal constraint) is kept high. Most of them are also elongated. Next, we combine these four representations.

## *12.3.3 Combining Balloons via Voting*

As we apply the decomposition method, each pixel in $C^o$ belongs to at least one balloon. These multiple representations result in many overlapping and redundant balloons, none of which (usually) captures a complete ground feature on its own. To merge the balloons and eliminate redundancy as much as possible, we apply a voting method mentioned in Sect. 6.1. Each pixel votes for the largest balloon of which it is a member. As votes are summed, those balloons receiving votes exceeding 80% of their areas are selected and retained. The rest are deleted as redundant

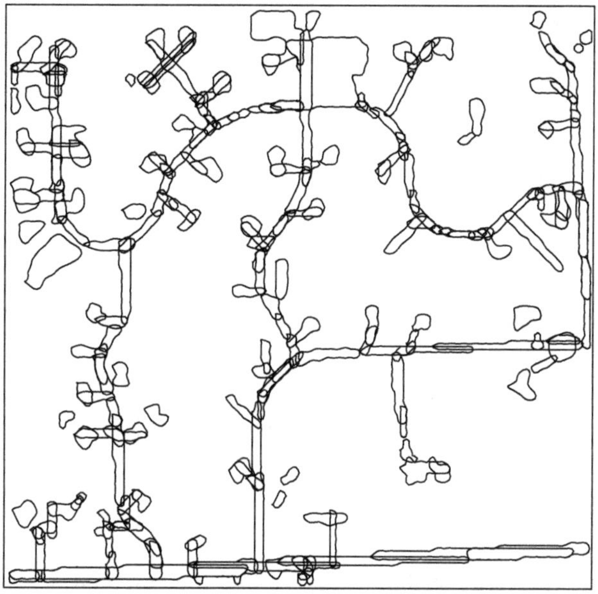

**Fig. 12.11** The South Dakota image, combination of balloons

since they do not convey new information. The merging does not result in a perfectly disjoint balloon set. The balloons may overlap somewhat around connecting regions; however, this overlap does not impair subsequent processing.

Applying the combination method to the initial balloons in Fig. 12.10, we obtain the final balloons shown in Fig. 12.11. This figure serves two purposes. First, it shows that the combination method is effective in eliminating most of the redundant balloons. Second, in obtaining elongated structures, our binary balloons work well even in such a complex environment. With the balloons in place, we are now prepared to abstract the description of the street network and houses, as discussed next.

## *12.3.4 Abstracting the Scene: Attributed Balloons*

To build an abstract representation of the scene, we attribute each balloon with the following properties: boundary, spine, length (of the spine), center of mass, and average width. We represent the outer boundary of each balloon using Fourier descriptors. Let a complex periodic function, $u(t) = x(t) + jy(t) = u(t + rT)$, $j = \sqrt{-1}$ for any integer values of $t$ and $r$, represent the outer boundary of the balloon. $T$ is the total number of points in the contour. The complex periodic contour can be approximated by a Fourier series as given in Chap. 6. In this chapter, we use the fourth order Fourier series representation to filter the boundary shape prior to computing its features below.

To obtain the spine of the balloon we use its curvature. The curvature is a differential geometric entity giving a measure of how rapidly the curve deviates

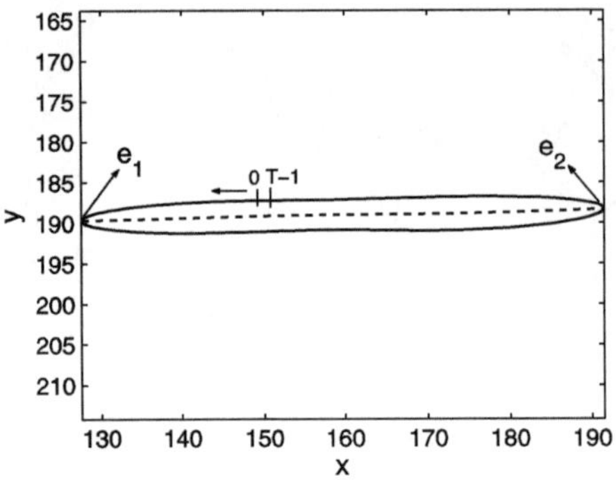


(a) Filtered version of the boundary and the spine extracted. solid curve - boundary, dashed curve - spine

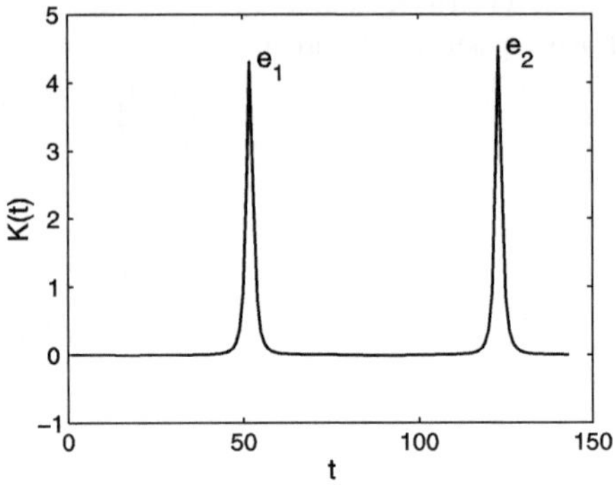


(b) The curvature of the boundary

**Fig. 12.12** Spine extraction

from its tangent line [12]. We find the curvature of the filtered boundary ($\hat{u}(t) = \hat{x}(t) + j\hat{y}(t)$) as

$$K(t) = \frac{\frac{d\hat{x}(t)}{dt}\frac{d^2\hat{y}(t)}{dt^2} - \frac{d\hat{y}(t)}{dt}\frac{d^2\hat{x}(t)}{dt^2}}{\left((\frac{d\hat{x}(t)}{dt})^2 + (\frac{d\hat{y}(t)}{dt})^2\right)^{3/2}}. \tag{12.5}$$

Given the balloon construction and filtering above, the extremal points of this curvature correspond to the endpoints, $e_1$ and $e_2$, of the spine (on the boundary) to be extracted. There is no algebraic solution to obtain the roots of (12.5) directly; we solve it numerically.

With $e_1$ and $e_2$ corresponding to the curvature extreme (spine endpoints) on the contour, we split the contour function into two parts at these points. These are arbitrarily labeled the upper and lower contours, $x_u + jy_u$ and $x_l + jy_l$, respectively. Both are reindexed to run from $e_1$ to $e_2$ and interpolated to have the same number of points. Then the spine of the balloon is defined as

$$s(k) = \frac{1}{2}\left[\left(x_u(k) + x_l(k)\right) + j\left(y_u(k) + y_l(k)\right)\right], \tag{12.6}$$

for $k = [0, e_2 - e_1)$.

To demonstrate the construction of the spine of a balloon, let us consider a horizontal balloon from the South Dakota image. We first obtain the filtered version of the boundary by a fourth order fit as in Fig. 12.12(a). We give the corresponding curvature in Fig. 12.12(b). And we find the extremal points of this curvature ($e_1, e_2$ labeled on both the boundary and the curvature). As can be seen, the extremal points of the curvature are distinct and easy to extract. We finally obtain the spine curve (represented by a dashed curve) in Fig. 12.12(a).

We keep the boundary and the spine in parametric form for further processing (such as street extraction). We take the length of the spine to be its arc length. The number of pixels on the spine could also be considered to be its length. However, the arc length is robust to non-uniform pixel placements along the spine. If the spine is given as $s(k) = x(k) + jy(k)$, then its arc length is

$$l = \int_{e_1}^{e_2} \sqrt{\left(\frac{dx(k)}{dk}\right)^2 + \left(\frac{dy(k)}{dk}\right)^2}\,dk. \tag{12.7}$$

The center of mass of each balloon is the centroid of its point set. To compute the average width of the balloon, we compute the distance from each contour point to the spine along the direction orthogonal to the spine. The average of these distances is taken as the width of the balloon.

## 12.4 Street Network and House Detection

This section describes how we extract the street network and houses from the balloons. We first eliminate balloons showing neither street nor house characteristics. Then, we construct a graph over the remaining balloons. Using graph-theoretical techniques and size information as discussed below, we jointly extract street networks and houses.

### 12.4.1 Eliminating Balloons Corresponding to Large Structures

In residential regions, we may also encounter large buildings such as shopping malls and schools, along with their parking lots. Therefore, some balloons produced in the previous step may represent these structures. We identify and discard these balloons by their morphology.

To discard balloons representing possible large buildings and parking lots, we use their area and compactness. Compactness is defined as the ratio of a region's area to the square of its perimeter, normalized by $4\pi$ [8].

Ours are metric images with the same resolution on the ground (4 meters/pixel multispectral) over the entire image set; so area is a reliable feature. A typical house is far smaller than a shopping mall or a parking lot. Road networks encompass large areas, of course, but they are highly non-compact. Therefore, we calculate compactness as a second feature. Balloons possibly belonging to the street network have low compactness values; balloons representing a house are highly compact and of small area. We eliminate balloons having *area* $> 75$ and *compactness* $> 4\pi/60$ as being neither part of the street network nor a house.

### 12.4.2 Forming a Graph to Represent the Balloon Neighborhoods

We extract the street network and houses using methods rooted in graph theory. For our previous methods, we also benefit from graph theory in Chap. 8. In this chapter, we follow the same notation for graphs.

We construct the balloon graph as follows. Each balloon is associated with a vertex in the graph. Therefore, each vertex $v_i$ has three properties: arc length $l(v_i)$, width $w(v_i)$, and the center of mass $c(v_i) = (x_i, y_i)$. There is an edge between two vertices if their corresponding balloons have a common boundary section. The weight assigned to this edge is the distance between these two balloons. To calculate this distance let $v_i$ and $v_j$ be two adjacent balloons having a common boundary. Let the center of mass of the common boundary be $c_c = (x_c, y_c)$. The distance between these two balloons (the weight of the edge connecting them) is

$$w(i, j) = \left\| c(v_i) - c_c \right\| + \left\| c_c - c(v_j) \right\|. \tag{12.8}$$

Since we use distance as a weight and will compute the shortest paths to recover the street segments, it is reasonable to use this definition. This weight definition approximates the distance between the two adjacent vertices along the spines of the corresponding balloons from one centroid to the next. Next, we consider the distance between two (not necessarily adjacent) vertices.

Berge [13] defines the distance $d(i, j)$ between two (not necessarily adjacent) vertices $(i, j)$ in a weighted graph to be the length of the shortest path from vertex $i$ to vertex $j$.

**Theorem 12.1** (Berge) *$d(i, j)$ satisfies*

1. $d(i, i) = 0$,
2. $d(i, j) + d(j, k) \geq d(i, k)$.

*In addition, if the graph is symmetric, we have*

3. $d(i, j) = d(j, i)$.

*This function is therefore a true metric in the topological sense.*

We obtain the distance between any two vertices by means of Dijkstra's shortest path algorithm given in Algorithm 5 [14]. This algorithm is one of the standard shortest paths algorithms used in graph optimization problems. It can also be implemented in parallel, since at each step it finds the shortest path between any two given vertices.

### 12.4.3 The Detection Step

Now, we turn our attention to street network and house detection, introducing a novel algorithm based on graph theory. With this algorithm, we first extract the

**Algorithm 5**: Dijkstra's shortest path algorithm

Begin
$w(a, b)$ = distance between two neighbor vertices $a$ and $b$
$X$ = vertex set of a graph $G$; $X = 1, 2, \ldots, m$
$s$ = initial vertex
$t$ = final vertex
$d(s, x)$ = the shortest path between vertex $s$ and vertex $x$, $x \in X$
$d(s, s) := 0$
$d(s, x) := \infty \ \forall x \neq s; x \in X$
$y$ = last vertex that was labeled
$y := s$
**while** (($d(s, t)$ has not been found) or ($d(s, t) \neq \infty$)) **do**
  **for** each unlabeled vertex $x$ **do**
    $d(s, x) := \min\{d(s, x), d(s, y) + w(y, x)\}$
  **end for**
  **if** $d(s, x) = \infty$ for all unlabeled vertices **then**
    $d(s, t) = \infty$;
  **end if**
  $y := \arg\min_x (d(s, x))$
**end while**
return $d(s, t)$
End.

street network (via balloons) and label the remaining balloons as houses. To extract the street network, we have unary and binary constraints. Unary constraints are used to detect balloons that could represent a street segment by themselves. Binary constraints are used to lace together balloons that could represent a street segment if considered in conjunction with their neighbors.

We assume that a balloon represents either a part of a house or a street segment at this point, based on the method and constraints of their construction. There are general house and street characteristics (at least for North America) that lead to a street network and house detection algorithm from the balloon graph. These are as follows.

- In residential regions, houses are connected to the street network via driveways. Since we take driveways as belonging to the street network, houses are located at its endpoints. In some cases, short, wide driveways may be assigned to the street network. This is not a major problem.
- Compared to an elongated street segment, the aspect ratio of a house is small. Therefore, a ratio threshold can help in discriminating house and street balloons to a great extent.
- A street network is topologically connected, with straight and curved sections. Straight sections produce fewer, longer balloons; curved sections produce more, shorter balloons.

We apply these observations to obtain the street network and houses from the balloon graph. For each vertex (balloon), we calculate its aspect ratio. This ratio will indicate the likelihood of that balloon belonging to a street network. We take balloons having a ratio greater than 7.5 to be street segments. This unary constraint initiates street segment extraction.

We then consider binary constraints to complete the extraction of the street network. We first discard any vertex of aspect ratio smaller than $t_r$ ($t_r = 5$, here) and having at most one neighbor. As we mentioned, such vertices are more likely to represent a house than a street segment. For vertices (balloons) having more than two adjacent vertices to belong to a street segment, their combination should exhibit the geometric characteristics of a street. Therefore, we retain paths of balloons (each member has at least two neighbors and an aspect ratio of at least five), with distances between their furthest vertices larger than a threshold $t_d$ (taken as five pixels, or 20 meters here). The underlying assumption is that a path of more than 20 meters length is more likely to be a street segment than a house group. In curved sections of the street network, we will have many balloons with more than one neighbor and aspect ratio less than five, but by chaining them together we can extract the street(s). Balloons not assigned to the street network are labeled as houses.

This algorithm may seem (too) simple; however, it exists in the context of a larger system. We have restricted our choices to a street segment or a house by earlier processing. We have similarly restricted the region under analysis to be residential. Such regions have relatively strict rules in terms of house and street locations. Houses and street networks lie in close proximity, but a street cannot pass through a house. Therefore, this relatively simple graph theoretical street network extraction method works well. The street network and house detection algorithm is given in Algorithm 6.

### *12.4.4 Road Tracking by Prediction*

Some street segments may be occluded by overhanging trees, be labeled as houses, or be eliminated as too small. To recover these missing street segments, we apply a prediction based road tracking method. Our method is similar to the correlation based road tracking methods summarized in [15].

We apply road tracking by prediction only to those street balloons having a single neighbor ($deg(v_i) = 1$). Therefore, we track possible streets only from endpoints of the balloon graph. We recall the parametric representation of their spines and for each spine representation we apply the following method.

Let $r(t) = (x(t), y(t))$ be the spine of a balloon with $deg(v_i) = 1$. It can be represented by a $k$th order polynomial on $n$ points:

$$x(t) = \sum_{i=0}^{k} a_i t^i, \quad y(t) = \sum_{i=0}^{k} b_i t^i \tag{12.9}$$

with $t = [1, 2, \ldots, n]$. Coefficients $a_i$ and $b_i$ are obtained from the least squares fit; we take $k$ to be one. The number of interpolated points $n$ is set to the least common multiple of the number of points in each half-contour.

**Algorithm 6**: Street network and house detection algorithm

Begin
*aspect ratio*$(v_i) := l(vi)/w(vi)$ such that $v_i \in V$
$t_r$ = threshold for aspect ratio; $t_r := 5$
$t_d$ = threshold for distance; $t_d := 5$
$R$ = street vertices; $R := \emptyset$
$H$ = house vertices; $H := \emptyset$
**for** $v := v_1$ to $v_m$ **do**
  **if** *ratio*$(v) > 1.5 \times t_r$ **then**
    $R := R \cup \{v\}$ (Unary constraint)
  **end if**
**end for**
$p(v_i, v_j)$ = path connecting $v_i$ and $v_j$; set of vertices
$G_1 := G \setminus v_i \ \forall\ v_i$ such that *ratio*$(v_i) < t_r$ and *deg*$(v_i) < 2$
$G_1 = (V_1, E_1)$; $V_1 = v_1, v_2, \dots, v_k$
**for** $i = 1$ to $k$ **do**
  **for** $j = (i+1)$ to $k$ **do**
    **if** $d(v_i, v_j) \geq t_d$ **then**
      $R := R \cup p(v_i, v_j)$ (Binary constraint)
    **end if**
  **end for**
**end for**
$H := V \setminus R$
return H, R
End.

Our prediction of the next point is $\hat{r}_1(n+1) = (\hat{x}(n+1), \hat{y}(n+1))$, extrapolating the polynomial fit. We also pick two neighbors of $\hat{r}_1(n+1)$, lying in the directions orthogonal to the spine curve, tagged $\hat{r}_2(n+1)$ and $\hat{r}_3(n+1)$. The next street point chosen is then $\hat{r}(n+1) = \arg\max_{\hat{r} \in \{\hat{r}_1, \hat{r}_2, \hat{r}_3\}} \Omega\{\hat{r}\}$ where $\Omega$ is the index of human activity (see (12.1)). We then fit a polynomial to the most recent $n$ points, and repeat the prediction. Tracking continues until $\max_{\hat{r} \in \{\hat{r}_1, \hat{r}_2, \hat{r}_3\}} \Omega\{\hat{r}\}$ drops below a threshold $t_p$, discussed next.

We want road tracking to continue only for pixels having sufficiently high $\Omega$. Therefore, we calculate $t_p$ in a Bayesian decision framework between object (high $\Omega$ values, $C^o$) and background (low $\Omega$ values, $C^b$) classes for each image separately ($t_p$ is therefore adaptive). We first obtain the conditional sample $\Omega$ distributions for object and background classes. We then set $t_p$ equal to the optimal Bayesian decision boundary value between these two classes. Rarely, this threshold may not be sufficiently strong to stop iteration. To handle these cases, we insert a control, checking the length of the extracted street segment. If the length exceeds the sum of the image width and height, we discard that prediction block. We assume that the prediction should encounter another road or the image perimeter within that distance. This constraint rarely comes into play.

We illustrate this method on the South Dakota image in Fig. 12.13. Although most of the street segment balloons have been labeled correctly, some were discarded while combining balloons and applying size constraints. One of the actual

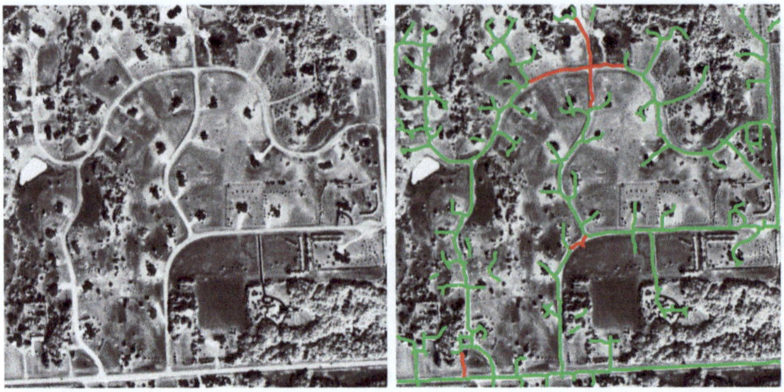

(a) The South Dakota image (b) The final street network

**Fig. 12.13** The South Dakota image, street network extraction. *Green sections* are obtained by the graph theoretical method; *red sections* are obtained by street tracking

street segment balloons is also labeled as a house. To correct these errors, we apply the road tracking algorithm. Figure 12.13(b) presents the street network extracted before tracking in green. The street network as extended by the road tracking algorithm is given in red. As can be seen, almost all missing street segments have been recovered by the tracking method.

### *12.4.5 Summary of System Parameters*

Table 12.2 summarizes the system parameters and how they are set. Eleven out of 13 parameters are fixed, as explained above, while two are adaptive (ratios). This number of parameters is not excessive when one considers the complexity of the task, and given that the system comprises several modules. As we have pointed out, all of these parameters have been carefully set in accordance with the metric and normalized nature of the images we consider, and none are particularly sensitive.

## 12.5 Results and Discussion

We tested our methods on 44 residential Ikonos images (each panchromatic image being 800 × 800 pixels, with 200 × 200 multispectral representations) taken from different locations around North America.

We first tested KMC-SC to extract possible house and street network pixels. Our house and street detection methods depend directly on this initial step. Next, we evaluated the house and street network detection steps separately. We offer four examples to show the system performance for four different residential region types.

**Table 12.2** Summary of system parameters

| System level | Parameter | Value |
|---|---|---|
| Eliminating lakes | region size | $> 75$ pixels |
| Eliminating lakes | median of $\theta$ | $< 0.2$ |
| Eliminating lakes | median of $\gamma_2$ | $< 0.3$ |
| Segmenting the $\Omega$ image | initial threshold $t_s$ | $< 0.8$ |
| Constructing binary balloons | overlap threshold $t_o$ | 50% of initial balloon size |
| Combining balloons | voting percentage | 80% |
| Abstracting the scene | Fourier series fit order | 4 |
| Eliminating balloons | area | $> 75$ |
| Eliminating balloons | compactness | $< 4\pi/60$ |
| Detection step | aspect ratio $t_r$ (street) | $> 7.5$ |
| Detection step | distance threshold $t_d$ | 5 pixels |
| Road tracking by prediction | fit level | 1 |
| Road tracking by prediction | tracking threshold $t_p$ | adaptive, Bayesian |

## *12.5.1 Pixel Based Classification*

In classification, we take house or street pixels ($C^o$) as one group and the background pixels ($C^b$) as another group. We have $44 \times 200 \times 200$ pixels since we are using the multispectral representation. We obtain 97.8% and 94.5% correct classification rates for the background and house or street pixels with an average classification performance of **97.3%**. The errors in this section are mostly due to overhanging trees.

## *12.5.2 House Detection*

Our image set includes a total of 6803 houses. Depending on the region, houses have different characteristics. Size, shape, and the setback from the street all vary, as does the spacing between them. To evaluate house detection performance, we report probability of detection $P_d$ and branching factor $B_f$, as used in Table 11.2. Here, we assume that a house is detected if any part of it is detected as in [16]. Because of the way we form overlapped balloons, very small (one or two pixel) house detections are virtually impossible. So we consider this approach to scoring the results to be reasonable. We obtain $\mathbf{P_d = 92.9\%}$ with $\mathbf{B_f = 9.5\%}$ under these circumstances, confirming the effectiveness of our house detection methodology.

### 12.5.3 Street Network Detection

In our 44 residential test images, the total length of street networks is 55442 pixels ($55442 \times 4$ meters). Similar to Table 11.4, we provide $P_d$ and $P_f$; we define $P_d$ in terms of length. In this section, we also evaluate the quality of road tracking by the prediction technique.

The street detection algorithm achieved $P_d = 89.9\%$ with $P_f = 3.8\%$ before prediction. Road tracking increased both: $\mathbf{P_d = 94.8\%}$, $\mathbf{P_f = 8.0\%}$. It improved $P_d$ by 4.9%, at the cost of roughly doubling $P_f$. Although somewhat application dependent, we consider the improvement in detection worth the increase in false alarm rate, at these levels. Therefore, tracking by prediction is an integral part of the system.

### 12.5.4 Some Detection Examples

We present house and street network detection results for four images (Indiana, Maryland, South Dakota, and South Dakota II). To demonstrate house and street network detection performance, we followed the same color code for both (green corresponds to a correct detection, blue corresponds to a miss, and red corresponds to a false alarm in subfigures (b) and (c)). We also provided the overall detection for each image (in subfigures (d)). Driveways are ignored in scoring because they could be considered part of the street network, part of a house, or neither. We have not attempted to resolve driveways as distinct entities.

The first result is for the Indiana image (Fig. 12.14). This residential region represents three distinctive characteristics. It has a mature region with well-spaced houses and trees nearby. There is a lake in the middle of the scene. The region in the upper left is a construction zone with new houses. Therefore, this scene is one of the hardest to process. There are 170 houses in the scene of which 165 are correctly localized with 12 false alarms. False alarms originate mostly from the spacing between houses and street segments. Some locations near street segments are falsely recognized as houses. Houses in the mature area are difficult to detect even for a human observer; nevertheless, the system was able to locate most of them. The 1525 pixel length street network is extracted except for a 10 pixel section, with a 90 pixel false alarm. Most false alarms occurred in the road tracking step, extending driveways for one or two extra pixels owing to the dense configuration of houses and street segments. The missing street segment could not be recovered by road tracking, since some portions of it were completely obscured by overhanging trees. The remaining street network is correctly detected with high accuracy for this scene.

The Maryland image (Fig. 12.15) depicts another type of residential region. It presents well-spaced and similar houses. Although there are trees in the scene, they are not close to the houses. This indicates that the neighborhood is newly constructed in a previously forested region. There are 151 houses in the scene, of which 145 are correctly located, with three false alarms. The well-spaced houses in this

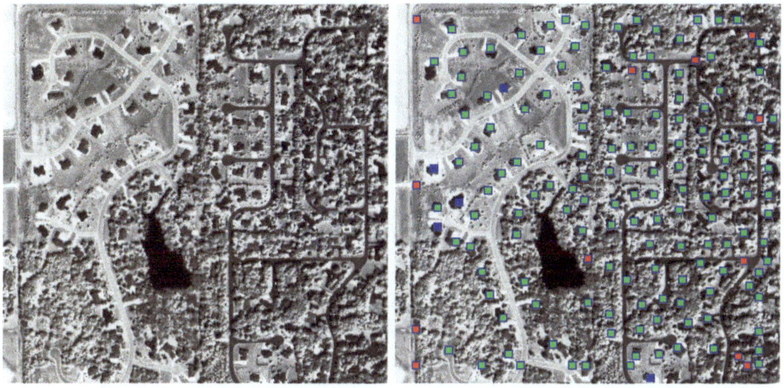

(a) The panchromatic image

(b) House detection; green - correct, blue - miss, red - false alarm

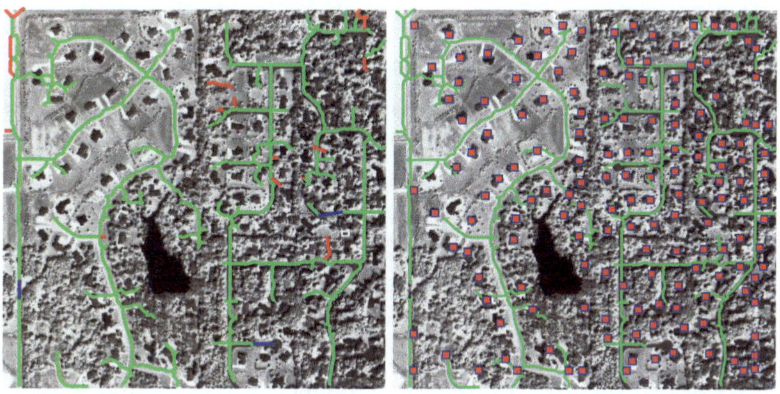

(c) Street network detection, green - correct, blue - miss, red - false alarm

(d) The overall detection, green - street network, red - house

**Fig. 12.14** The Indiana image, street network and house detection

scene give rise to a low false alarm rate. For this image, the entire 1133 pixel street network is extracted but for a 46 pixel section. The majority of the missing street section is near the top, occluded by trees.

The third result is for the South Dakota image (Fig. 12.16). This residential region represents low density housing, with minor or no occlusions on the houses and streets. There is also little vegetation in the scene. Almost all houses and the street network can be seen clearly. This region is one of the easiest to process. In this residential region with 69 houses, 67 are correctly located with one false alarm. 1258 pixels of the 1276 pixel street network are detected correctly, with a 73 pixel false alarm. Although road tracking was able to recover most of the missing seg-

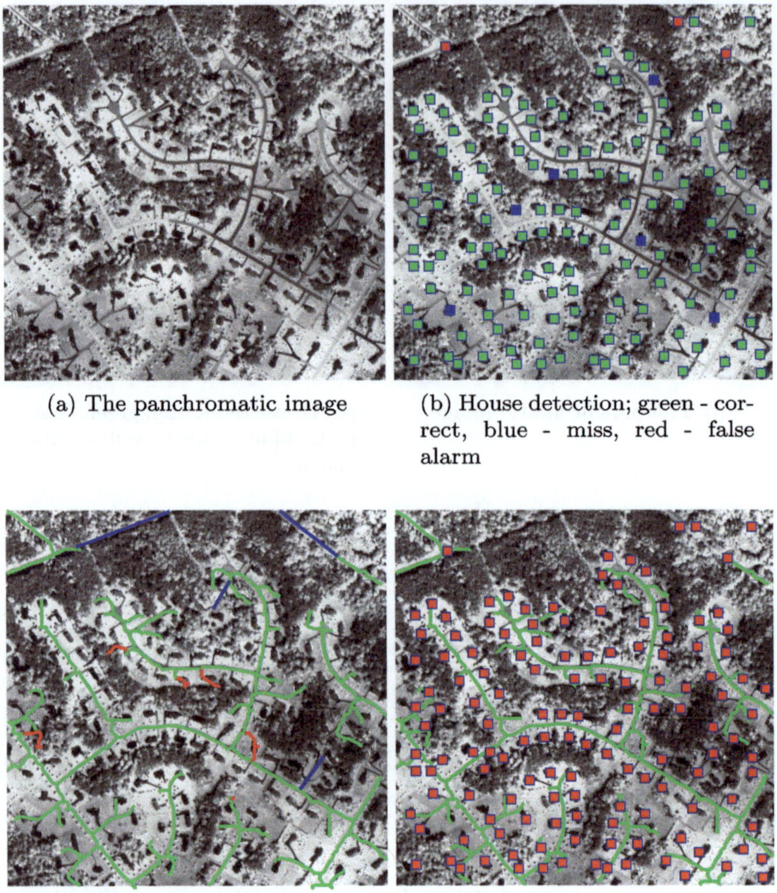

(a) The panchromatic image

(b) House detection; green - correct, blue - miss, red - false alarm

(c) Street network detection, green - correct, blue - miss, red - false alarm

(d) The overall detection, green - street network, red - house

**Fig. 12.15** The Maryland image, street network and house detection

ments for this image, it also increased the false alarm rate, mostly because of the low vegetation cover (causing extra one or two iterations).

The South Dakota II image (Fig. 12.17) is similar to the Indiana image. It is composed of both mature and newly constructed regions. Although there are fewer houses in the scene, the occlusion by overhanging trees is more prevalent than in the Indiana image. The spacing of the houses is similar to that of the Maryland image. For this region, 141 of 147 houses are correctly located, with six false alarms. The 1411 pixel length street network is detected with 193 pixels missing. Two main street segments are missing, the first being at the top of the scene, the second being near the middle. A street segment at the top left of the image was missed in the KMC-SC process. There is a heavy local occlusion that blocks the extension of the detected segment into this (wide, clear) region via the only possible path. Mean-

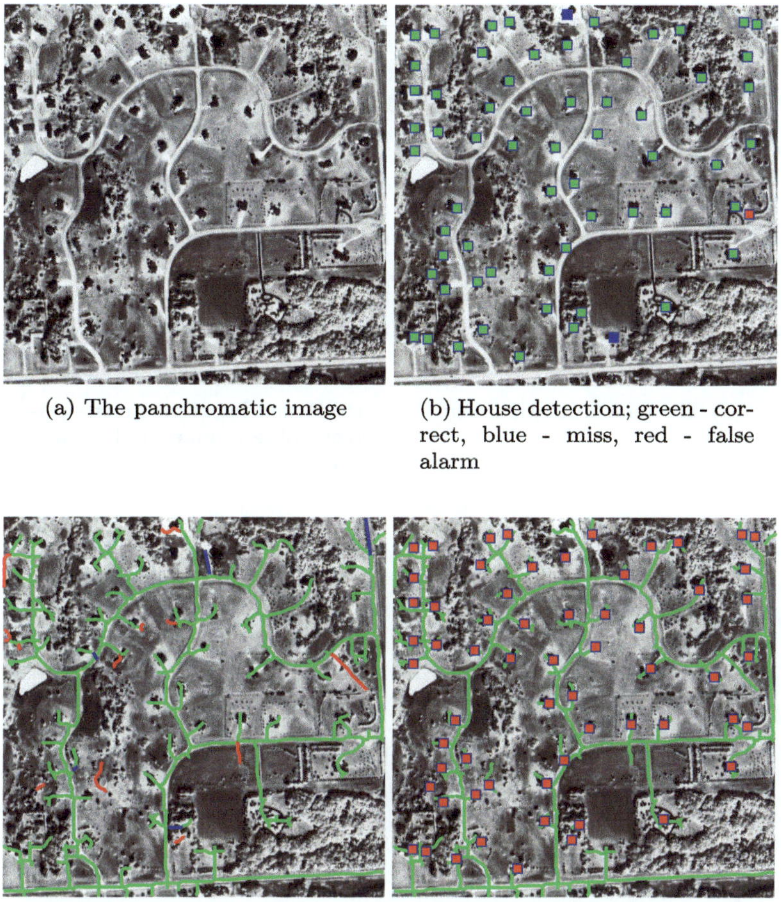

(a) The panchromatic image

(b) House detection; green - correct, blue - miss, red - false alarm

(c) Street network detection, green - correct, blue - miss, red - false alarm

(d) The overall detection, green - street network, red - house

**Fig. 12.16** The South Dakota image, street network and house detection

while, a smaller, partly obscured street section nearby is captured in the extension process because it connects to the (initially) detected network in four locations, any one of which would suffice. The false alarm count for street extraction in this image is 68 pixels.

As we observe these four results, we can comment on possible weaknesses of our system. First, as a region becomes more congested, the house and street network detection performance decreases. This is to be expected. Road tracking increases the false alarm rate considerably in street detection. Our detection system, without higher level perceptual reasoning, cannot overcome significant occlusion by trees. Therefore, if a house or a street segment is totally occluded by trees, it will not be detected. At the level of pixel classification, of course, this is not an error. This

(a) The panchromatic image

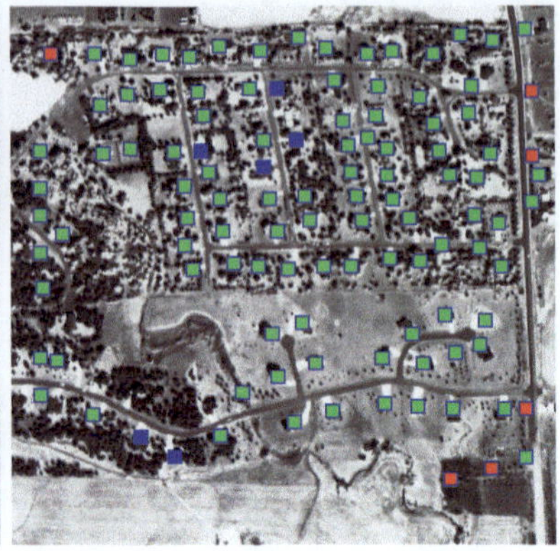

(b) House detection; green - correct, blue - miss, red - false alarm

(c) Street network detection, green - correct, blue - miss, red - false alarm

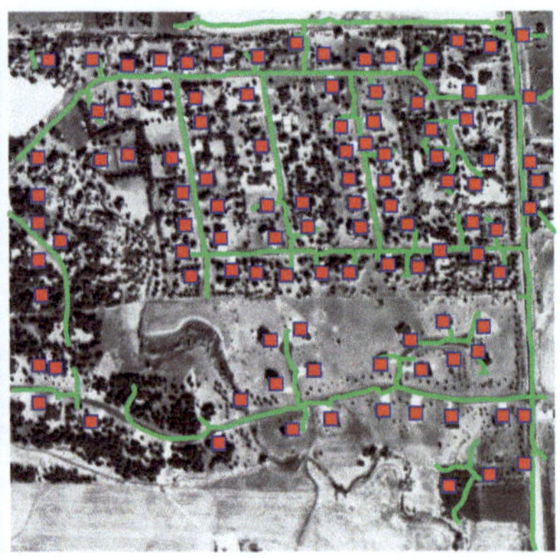

(d) The overall detection, green - street network, red - house

**Fig. 12.17** The South Dakota II image, street network and house detection

could be probably addressed by introducing a higher-level reasoning process to infer missing street segments. Even fairly simple collinearity or smooth continuation constraints may correct many of these errors [17].

These are the minor problems the system exhibits. On the other hand, these four different types of regions show the robustness of our house and street detection system to different conditions. In these four images, most of the houses and street network were detected correctly. For the Indiana and South Dakota II images, it is extremely difficult for a human observer to detect the houses in the mature (wooded) sections. Even in these regions, the system was able to detect and locate most houses correctly.

## 12.6 Summary of the Chapter

This chapter introduced a system to detect street networks and houses in residential regions via satellite images. The system comprises four parts: multispectral information processing, segmentation, decomposition, and graph theoretical analysis.

To locate possible human activity, we used the linearized vegetation index, $\theta$, which gives a low response to rock derivatives. Since these are the basic blocks for most of the houses and streets in residential regions, their presence indicates possible human activity. And this was the case for our test images. Because surface water responds similarly, we introduced a shadow–water index, and used the combination of these indices to eliminate water regions. We believe other applications exist for these indices.

Having successfully identified the areas of human activity, we introduced a modified k-means clustering algorithm to extract a binary segment representing possible houses, street networks, schools, malls, and parking lots. The novelty in our algorithm is the introduction of spatial coherence to clustering via connected components analysis.

To extract houses and street networks on this binary segment, we introduced a decomposition algorithm inspired by balloon algorithms. Our aim was to extract elongated structures (representing possible street segments) via this algorithm. To represent the binary image reliably (covering the curved regions, specifically), we have multiple balloon representations for the same region. We were able to overcome redundancy by applying a voting method. This overall scheme worked well.

We represented these balloons with a weighted a graph to extract street networks and houses. At this step, we eliminated balloons representing neither a street nor a house structure. Therefore, at this step, a balloon in the graph is either representing a house or a street segment. By invoking simple unary and binary constraints, we were able to detect houses and street networks.

We finally tested our system with 44 residential images. Our performance on such a diverse and large test set is noteworthy. The overall system may ultimately find use as an automatic map generation system specialized for residential regions.

## 12.7 Problems

**12.1** How do vegetation indices help when detecting buildings and street segments in residential regions?

**12.2** How can shadow and water pixels be separated in the SWI image?

**12.3** What is the main difference between KMC and KMC with spatial coherence?

**12.4** What basic assumption does Otsu's thresholding method depend on?

**12.5** Compare the standard balloon algorithm with the binary balloon algorithm introduced in this chapter.

**12.6** How can the method introduced in this chapter be labeled based on the grouping in the previous chapter?

**12.7** (**Open ended question**) Can there be more unary and binary constraints embedded into the graph in detecting buildings and street segments?

## References


1. T.E. Avery, G.L. Berlin, *Fundamentals of Remote Sensing and Airphoto Interpretation*, 5th edn. (Macmillan Publishing, Indianapolis, 1992)
2. R.O. Duda, P.E. Hart, D.G. Stork, *Pattern Classification*, 2nd edn. (Wiley Interscience, New York, 2001)
3. C. Ünsalan, K.L. Boyer, Comput. Vis. Image Underst. **98**(3), 432 (2005)
4. J.A. Hartigan, *Clustering Algorithms* (Wiley, New York, 1975)
5. A.K. Jain, R.P.W. Duin, J. Mao, IEEE Trans. Pattern Anal. Mach. Intell. **22**(1), 4 (2000)
6. N. Otsu, IEEE Trans. Syst. Man Cybern. **9**(1), 62 (1979)
7. R. Adams, L. Bischof, IEEE Trans. Pattern Anal. Mach. Intell. **16**(6), 641 (1994)
8. C. Godsil, G. Boyle, *Algebraic Graph Theory* (Springer, New York, 2001)
9. M. Kass, A. Witkin, D. Terzopoulos, Inter. J. Comput. Vis., 321–331 (1988)
10. R.C. Gonzales, R.E. Woods, *Digital Image Processing* (Addison Wesley, Mass., 1993)
11. R.A. Brualdi, *Introductory Combinatorics*, 3rd edn. (Prentice Hall, New Jersey, 1999)
12. M.P. Do Carmo, *Differential Geometry of Curves and Surfaces* (Prentice Hall, New Jersey, 1976)
13. C. Berge, *The Theory of Graphs* (Dover, New York, 2001)
14. J.R. Evans, E. Minieka, *Optimization Algorithms for Networks and Graphs*, 2nd edn. (Dekker Inc., New York, 1992)
15. D.M. McKeown, J.L. Denlinger, in *Proceedings of CVPR1988* (1988), pp. 662–672
16. C. Lin, R. Nevatia, Comput. Vis. Image Underst. **72**(2), 101 (1998)
17. E. Riseman, X. Wang, B. Hanechak, H. Schultz, A. Hanson, in *Fifth International Conference on Computer Graphics and Visualization*, 1997

# Part VI
# Summarizing the Overall System

# Chapter 13
# Final Comments

In this book, we proposed an end-to-end multispectral satellite image understanding system. Our system starts with land use classification and ends with house and street network detection in residential regions. In this final chapter, we summarize what has been done in each chapter and emphasize important observations.

In the first part of the book, we summarized the remote sensing satellite families as well as their properties. Based on the launching dates of these satellites, we can infer that more and more remote sensing satellites with improved resolution and spectral ranges will be available in the future. Therefore, the need for more intelligent and fast automated systems to analyze the images from these satellites will be enormous.

In the second part of the book, we examined vegetation and shadow–water indices as possible methods for summarizing the multispectral information. We considered the best known vegetation index in the remote sensing literature, *NDVI*, and developed a rigorous statistical justification for it by using the principal components analysis. As we observed, the *NDVI*, in fact, corresponds to a slope in the statistically decorrelated vector space. Armed with this interpretation, we then proposed a solution to the saturation problem of the *NDVI*. Simulations and experiments on real images show that our proposed measure overcomes the saturation problem.

Next, we considered the relationship between the *NDVI* and the *RVI* and showed that these two indices, in fact, capture the same information. With the statistical framework at hand, we considered more bands and introduced new vegetation and shadow–water indices. We compared these new vegetation indices with the angle measure we derived from the *NDVI* on real images. We conclude that while all of the indices introduced can be used to detect vegetation successfully none clearly outperforms $\theta$, based on the original *NDVI* ratio. We also compared shadow–water indices on real images and found that $\gamma_2$ based on the blue, red and near-infrared bands works best.

The third part of the book was about land use classification. We first reviewed the literature to observe the trends and performances in land use classification. Since, most of the papers reported performances over a single image, we could not reach a conclusion on their relative performances. However, we observed that none of these

C. Ünsalan, K.L. Boyer, *Multispectral Satellite Image Understanding*,
Advances in Computer Vision and Pattern Recognition,
DOI 10.1007/978-0-85729-667-2_13, © Springer-Verlag London Limited 2011

studies were based on the structural properties of the panchromatic images which we believe is the most promising direction for the publicly-available high resolution satellite images.

To that end, we first presented a new approach to assess the degree of land development. While spectral signatures have been exploited for years in land use classification, our objective was to learn what information regarding land development could be efficiently extracted from the photometric *structure* in the image. We based the assessment on the photometric and geometric characteristics of straight line segments. Although edge detection may be used to extract lines, we used a more robust *straight line* extraction method based on regions of consistent gradient orientation.

Among several features we developed, those based on length and contrast proved to be the most promising for classification. The strength of these features is that they do not depend heavily on the city model; they apply to most cities around the world, particularly over $400 \times 400$ m$^2$ windows. The weaker features are those more dependent on a specific city model. The model assumes a high(er) degree of organization, which may not hold for many old world cities. However, if one's focus were on urban and rural discrimination in North America (and Australia, for example), then these features may prove more useful.

We tested both parametric (Bayes) and non-parametric (Parzen window and nearest-neighbor) classifiers. Although these classifiers differ in their assumptions with respect to the underlying data distribution, they offer similar performance on our data. However, once trained, the Bayes classifier requires the least computation; the nearest neighbor the most. Therefore, we would recommend the Bayes classifier for a production system using this type of data. With probabilistic relaxation, we obtained slightly improved classification performance. However, this improvement is marginal considering its additional computational cost, and we would not recommend it.

With the help of this multispectral information, we improved our previous results in discriminating urban and non-urban areas in three steps. Our first step was to include the linearized *NDVI* as multispectral information. Our second step was a synergistical combination of the structural and the multispectral information. Our last step was using the spatial information in classification. Our most promising features are those using the combination of structural and multispectral information. By using parametric and non-parametric classifiers, we were able to distinguish developed areas with high confidence.

Next, we introduced and tested several graph-theoretical measures to assess land development in overhead imagery. Our first objective was to develop measures that increase monotonically with respect to the organization (development) in images. The third measure on graph partitioning performed the best among others, in terms of false alarm rate. The main reason for this is the contrast information it has. Other contrast based structural features also proved to be useful as structural features. Although each measure performed fairly well alone, the measure obtained by fusing them performed best of all with respect to deviation (this measure matches the human opinion quite well). It captures different properties of the scene by different measures, and these seem to compensate each other.

Although our test set is large and diverse in location and development pattern, we can distinguish developed areas with high confidence. Our results indicate that image structure, as captured by the spatial organization of its straight lines, does provide an effective indicator of land development activity.

Our next objective was detecting sparse residential regions which are difficult to discriminate. In the fourth part of the book, we focused on this problem. The detection of suburban residential regions is highly challenging. Nevertheless, a novel feature based grouping method produced highly encouraging results. This suggests that methods based on principles from perceptual organization may ultimately have significant impact on these types of problems.

In terms of graph-theoretical measures, we first cast this as a one-class problem and tested it with 281 samples. We obtained a clear and definite improvement by using graph based features over structural ones. This indicates that for fine classification, structural features alone are not enough. Besides the structure in the scene, neighborhood information of these structures (captured by a graph here) are also needed.

Finally, we built a three-class classifier and tested it with 270 samples. For the three-class case, we obtained classification results superior to those reported anywhere over such a number of test samples. Again, this test indicates the necessity of graph-theoretical measures for fine classification. With these very high detection rates, our approach can be used to detect residential regions automatically.

In the fifth part of the book, we introduced a subsystem to detect street networks and houses in extracted residential regions. Our subsystem comprises four parts: multispectral information processing, segmentation, decomposition, and graph-theoretical analysis.

We started from the multispectral images to locate possible human activity. We used the properties of $\theta$ for this purpose. This index (and the *NDVI*, of course) responds low to rock derivatives. Since these are the basic blocks for most of the houses and streets in residential regions, their presence would indicate possible human activity. And this was the case for our test images. Unfortunately, lakes and shadows also respond like rocks to this index. To overcome this problem, we used the shadow–water index, and then eliminated only water regions. This human activity index, as well as water region extraction method can be used for applications other than ours.

Having successfully identified the areas of human activity, we introduced a modified k-means clustering algorithm to extract a binary segment representing possible houses, street networks, schools, malls, and their parking lots. The novelty in our algorithm is its introduction of spatial coherence to clustering via connected components analysis.

To extract houses and street networks on this binary segment, we introduced a decomposition algorithm inspired by balloon algorithms. Our aim was to extract elongated structures (representing possible street segments) via this algorithm. To represent the binary image reliably (covering the curved regions, specifically) we have multiple balloon representations for the same region. We were able to overcome redundancy by applying a voting method. This overall scheme worked fairly well in decomposing binary images we have.

Then, we represented these balloons in a graph framework to extract street networks and houses. At this step, we eliminated balloons representing neither a street nor a house structure. Therefore, at this step, a balloon in the graph is either representing a house or a street segment. By unary and binary constraints, we were able to detect houses and street networks on this graph.

We tested our house and street network detection system with 44 residential images. Our performance on such a diverse and large test set is noteworthy. We plan to use the spatial constraints (formalizing them as of either probabilistic relaxation or Bayesian networks) to refine our house and the street network labels. The overall system hence may be of use as an automatic map generation system focused on residential regions.

Finally, the overall system works fairly well. We justified this by the statistical performance of each step. This system may be used to generate maps labeling houses and the street network in residential regions, as well as locating these neighborhoods.

# Index



C. Ünsalan, K.L. Boyer, *Multispectral Satellite Image Understanding*,
Advances in Computer Vision and Pattern Recognition,
DOI 10.1007/978-0-85729-667-2, 

MIX
Papier aus verantwortungsvollen Quellen
Paper from responsible sources
FSC® C105338

If you have any concerns about our products,
you can contact us on
**ProductSafety@springernature.com**

In case Publisher is established outside the EU,
the EU authorized representative is:
**Springer Nature Customer Service Center GmbH**
**Europaplatz 3, 69115 Heidelberg, Germany**

Printed by Libri Plureos GmbH
in Hamburg, Germany